FNNR
Foundation for Neurofeedback
& Neuromodulation Research

Beginning Neurofeedback in Your Practice:
A Guide for Clinicians Using Neurofeedback From Intake to Discharge

Robert E. Longo, MRC, LCMHC, NCC, BCN

Becky Bingham, MS, RN, QEEG-D, BCN

Also by Robert Longo

Longo, R. E. (2020). My journey into neurofeedback. In J. R. Evans, M. B. Dellinger & H. L. Russell (Eds.), *Neurofeedback: The first fifty years.* Cambridge, MA: Academic Press/Elsevier, Inc.

Longo, R. E., & Soutar, R. (2019). *Becoming certified in neurofeedback.* Greenville, SC: FNNR.

Longo, R. E. (2018). *A consumer's guide to understanding QEEG brain mapping and neurofeedback training.* Bloomington, IN: iUniverse.

Longo, R. E., & Russo, M. (2017). Working with forensic populations: Incorporating peripheral biofeedback and brainwave biofeedback into your organization or practice. In T. F. Collura & J. A. Frederick. (Eds.), *Handbook of clinical QEEG and neurotherapy.* London: Routledge.

Longo, R. E. (2015). The use of neurofeedback to treat traumatic brain injury in sexually abusive youth. In D. S. Prescott & R. J. Wilson. (Eds.), *Very different voices: Perspectives and case studies in treating sexual aggression.* Holyoke, MA: NEARI Press.

Longo, R. E., Prescott, D. S., Bergman, J., & Creeden, K. (Eds.). (2013). *Current perspectives & applications in neurobiology: Working with young persons who are victims and perpetrators of sexual abuse.* Holyoke, MA: NEARI Press.

Longo, R. E., & Prescott, D. S. (2013). Current perspectives and applications in neurobiology: An overview. In R. E. Longo, D. S. Prescott, J. Bergman & K. Creeden (Eds.), *Current perspectives & applications in neurobiology: Working with young persons who are victims and perpetrators of sexual abuse.* Holyoke, MA: NEARI Press.

Longo, R. E. (2013). Traumatic brain injury (TBI): A brief introductory overview. In R. E. Longo, D. S. Prescott, J. Bergman & K. Creeden (Eds.), *Current perspectives & applications in neurobiology: Working with young persons who are victims and perpetrators of sexual abuse.* Holyoke, MA: NEARI Press.

Longo, R. E., & Prescott, D. S. (2013). Ethical responsibilities in neuroscience and neurobiology. In R. E. Longo, D. S. Prescott, J. Bergman & K. Creeden (Eds.), *Current perspectives & applications in neurobiology: Working with young persons who are victims and perpetrators of sexual abuse.* Holyoke, MA: NEARI Press.

Soutar, R. G., &; Longo, R. E. (2022). Doing neurofeedback: An introduction (2nd ed.). Foundation for Neurofeedback and Neuromodulation.

FNNR Publications

Longo, R. E., & Soutar, R. (2020). *Becoming certified in neurofeedback: A guide to the neurofeedback mentoring process for mentors and mentees.* Foundation for Neurofeedback and Neuromodulation Research.

Sokhadze, E., & Casanova, M. F. (Eds.). (2019). *Autism spectrum disorder: Neuromodulation, neurofeedback, and sensory integration approaches to research and treatment.* Foundation for Neurofeedback and Neuromodulation Research.

Martins-Mourao, A., & Kerson, C. (Eds.). (2017). *Alpha-theta neurofeedback in the 21st century: A handbook for clinicians and researchers* (2nd ed.). Foundation for Neurofeedback and Neuromodulation Research.

Soutar, R. G., & Longo, R. E. (2022). Doing neurofeedback: An introduction (2nd ed.). Foundation for Neurofeedback and Neuromodulation.

Hammond, D. C., & Gunkelman, J. (2011). *The art of artifacting.* The ISNR Research Foundation.

Carmichael, J. (2011). *Multi-component treatment for post-traumatic stress disorder, including strategies from clinical psycho-physiology and applied neuroscience.* The ISNR Research Foundation.

Donaldson, S. (2012). *The other side of the desk.* The ISNR Research Foundation.

Thompson, M., Thompson, J., & Wenqing, W. (2009). *ADD centre Brodmann areas booklet.* The ISNR Research Foundation.

Beginning Neurofeedback in Your Practice:
A Guide for Clinicians Using Neurofeedback From Intake to Discharge

Robert E. Longo, MRC, LCMHC, NCC, BCN

Becky Bingham, MS, RN, QEEG-D, BCN

The ISNR Learning Series

Beginning Neurofeedback in Your Practice: A Guide for Clinicians Using Neurofeedback From Intake to Discharge

Robert E. Longo, MRC, LCMHC, NCC, BCN

Becky Bingham, MS, RN, QEEG-D, BCN,

Publisher:

The FNNR
The Foundation for Neurofeedback & Neuromodulation Research 2131 Woodruff Rd, Ste 2100 #121
Greenville, SC 29607
http://www.theFNNR.org

Correspondence: admin@theFNNR.org

Publisher:

The FNNR

The Foundation for Neurofeedback & Neuromodulation Research 2131 Woodruff Rd, Ste 2100 #121 Greenville, SC 29607 http://www.theFNNR.org. Correspondence: Admin@ theFNNR.org

Layout Design: Megan Stevens. Correspondence: Publishing@ theFNNR.org

Printed and Bound by: BMED Press, LLC

Beginning Neurofeedback in Your Practice: A Guide for Clinicians Using Neurofeedback From Intake to Discharge

ISBN: 979-8-218-02075-0

Contents

Dedication

We can never fully thank all the clients, patients, and people in our lives who contribute to our knowledge and thus the writing of a book like this. We would first like to thank the ISNR Board of Directors for supporting and having the belief in us to create this learning series.

We would like to thank our many colleagues who have shared information along the way and helped us understand the challenges and difficulties they and others have faced. And of course, we would like to thank our clients who have helped us learn better and more effective methods and ways of working with them.

Preface

I came into the field in the late 1990s through the laboratory of Joel Lubar at the University of Tennessee in Knoxville. At that time the field was young despite having been established and having had practitioners for several decades. Yet information was rather sparsely distributed, and the clinical application of neurofeedback was less standardized and sometimes even considered a bit rogue. Clinical professionals remained a widely varied group of individuals ranging from mental health care professionals such as psychologists and counselors, marriage and family therapists, and social workers. It also included medical professionals such as physicians, nurses, physical therapists, and occupational therapists, although these later were far less common. Lucky for me, I was in a well-structured environment consisting of university laboratory experiences, including the teachings about the fundamentals of EEG data collection, analytics, processing, combined with clinical interpretation, and the application of operant conditioning of the EEG to treat a wide variety of clinical pathologies and symptom presentations.

The bulk of the individuals in the community of neurofeedback providers were trained in their clinical profession and then acquired neurofeedback skills or the ideas of applying neurofeedback to their clients later in their professional career. This was very challenging because the resources available at that time were more limited. Workshops and trainings were around but sparse. Those individuals often had to take weeks from work to travel to workshops and spend 4 to 5 days learning the fundamentals only to return to their practice where they were often alone, professionally isolated and trying to figure it out on their own.

Over the past two decades and more authors have been filling this void through both training and writings. Valuable resources from certification boards such as the Biofeedback Certification International Alliance (BCIA) and membership organizations such as The International Society for Neuromodulation and Research (ISNR) have provided a network and an opportunity for individuals to gather regularly both in person and virtually for trainings and the opportunity to mix and mingle among their colleagues, share stories, and seek clinical advice.

It was at an ISNR meeting in the mid 2000's that I met Robert Longo. Rob always struck me as an individual who was rising to top of his craft through dedication of study and application although he had previously been in other domains of mental health care and established quite extensive expertise in those domains. Yet he ventured into this world of neuromodulation and quickly became very active in the community with resounding success as a clinician, mentor and author. It was also at a future ISNR that I first met Becky Bingham. Becky was coming from the medical profession and had a slightly different yet very valuable perspective to neurofeedback. Collectively these two individuals have spent considerable time and effort in being involved in the community, preparing studious materials, consulting with a variety of individuals, and coalescing domain expertise from the different perspectives and different backgrounds of other neurofeedback professionals. It has been a pleasure to know and collaborate with both of them over the past years providing workshops at the aforementioned meetings, discussing mentorship or case consultations, and being partnered in community service to the profession and the group of providers that the professional organizations serve.

The ability of these authors to take very challenging concepts and distill them down to the most important and fundamental levels is a real skill. It is not that they take a complicated subject and simply leave things out in order to make it more comprehensible, they actually understand the content and have been able to then teach and describe these concepts in a meaningful way to new practitioners. Quite honestly, it's not uncommon that a very established individual who's been in the community for a long period of time will have a conversation and then candidly at some moment say, "Oh I actually didn't remember or perhaps didn't ever learn this concept." So, while this book is aimed at individuals who are newer to the profession and to the application of neurofeedback, it is appropriate for all providers to refresh their memories or to gain insight into what are perhaps industry standards of implementation. As I reflect on my first experiences in the clinical world, I wish I had this type of information readily available to me.

This book, Beginning Neurofeedback in Your Practice; A Guide for Clinicians Using Neurofeedback from Intake to Discharge, is an exceptional resource for providers to understand or begin to implement neurofeedback into their practice. The authors take concepts and encapsulate them in a meaningful and understandable way. Then they do what is often not taught at academic workshops or in university programs. They teach how to implement neurofeedback into a practice as a modality in a responsible, effective, ethical, and economically advantaged way for both client and provider.

This book begins with understanding why a clinician would want to introduce neurofeedback into their practice providing opportunities of mentoring, conferences, certifications and the resources that a provider would want to know in order to successfully set themself up for having a practice of neurofeedback. A very complicated issue not only because of the diversity of the providers who provide neurofeedback but also due to the claims and widely varying symptoms to which neurofeedback may be applied. The book moves into understanding branding and marketing avenues, touching on very complicated concepts such as: "How do we present ourselves to the community in a meaningful way? How do we market? What methods allow us to speak to the incredible power of the modality without being sensational and crossing lines of ethical or legal claims?" They thoughtfully explain how we can be ethical and responsible and still provide the potential client the opportunity to understand the full benefit of what neurofeedback may have to offer them.

The book then moves into more practical applications of how an intake may look. How we record assessment quantitative EEG and other tools, and the implications of the assessment process as a robust experience that is necessary for your clients. The authors then turn our attention to how to determine protocol selection and how to implement these protocols into the session. This is a very common theme that occurs in training workshops and didactic programs, where the clinician may understand the application and the mechanisms of action, and yet, not fully grasp what the practical steps may look like in their practice. In summary, Beginning Neurofeedback covers the basics from the very practical setup of where to place chairs and make individuals comfortable to how we care for our equipment and implement appropriate hygienic processes.

Finally, the last chapters guide the reader into the very practical components of keeping our clients informed of their progress, how we document and keep records, and lastly into closing and dismissing them. To emphasize the importance of these practical considerations, I have observed from some providers that some clients don't finish with intention. Instead, there is a long-drawn-out process until the client finally stops appearing. This is a less than ideal approach. It is important, just like the with other modalities that we employ as mental health professionals, that we understand how to appropriately terminate and provide ongoing support for clients as they leave our practice.

You'll find a set of essential information in the appendices of the book including manufacturers of equipment, sample forms such as progress notes, and an essential skills list that is not only consistent with certification agencies but also one that any provider would want to make sure that they have mastered as they begin to incorporate neurofeedback in their practice. The authors have not only provided an exceptional resource for new providers who are exploring neurofeedback but also for the experienced provider.

The authors have the domain expertise, the background experience and they have, most importantly, put the work in themselves of providing neurofeedback to clients and establishing a successful practice. I can't imagine any individuals who could be more qualified or more accessible in providing this information than Rob Longo and Becky Bingham.

In summary, it is an honor to have been able to read a sneak peak of this book. I found it valuable to my own practice, but I am most excited for the reader to be able to have this information as an early adopter rather than having to explore through trial and error on their own. When I was asked if I would be willing to write this forward, I knew that I would want the opportunity to communicate and endorse the efforts of these authors for our community and for you, the clinicians who are implementing these valuable skills into your practice. I have no doubt that you will find the information valuable, whether you are a novice or a well experienced practitioner. This writing is a great service to the field, and I hope that you enjoy it as much as I have. My deep appreciation to the authors for their time, effort and contribution to the field. May you all have overwhelming success with your clients. I have no doubt you will truly be amazed at the experiences that your clients will report back to you from using neurofeedback in your practice. My very best wishes.

-Leslie Sherlin, PhD, MAC, MSc, LAC, CMPC, NCC, BCN, BCB, ECP, QEEGD

Certified Sport and Performance Consultant & Licensed Counselor

Private Practice, Sherlin Consulting Group, Scottsdale Arizona

Professor of Psychology, Ottawa University, Surprise Arizona

Senior Doctoral Adjunct Dissertation Chair, Grand Canyon University, Phoenix Arizona

Associate Professor, Southwest College of Naturopathic Medicine & Health Sciences, Tempe Arizona

Introduction

Introducing Neurofeedback into Your Current or New Practice

Neurofeedback is an exciting and beneficial tool for working with disorders of the brain and central nervous system such as attention deficit hyperactive disorder (ADHD), anxiety, depression, post traumatic distress disorder (PTSD), and autism spectrum disorder (ASD) to name a few. A growing number of professionals are adding neurofeedback to their practice or starting a business solely to offer neurofeedback.

There are many things you will need to do to make sure you become competent, your sessions run smoothly, and that your clients are getting the best care possible.

Since neurofeedback is a highly specialized intervention, you will need specialized training and mentoring, specific equipment and software information, supplies, and clinical forms, in order to properly provide these services. Whether you are adding neurofeedback to an existing practice or starting a new business, learning neurofeedback can often feel challenging and overwhelming.

This book is designed to guide you through the process of obtaining training and setting up your neurofeedback practice. It will help ensure that the introduction of neurofeedback into your practice is smooth and efficient. As seasoned clinicians, we have spent over 30 years collectively providing neurofeedback services to our clients, and we have worked in a variety of settings from psychiatric hospitals, residential treatment centers, medical hospitals, clinics, medical offices, and private practice.

We encourage you to join organizations that have

a focus on neurofeedback and offer workshops on adding neurofeedback to your practice, marketing, strategies, etc., such as the *International Society for Neuroregulation and Research (ISNR), the Association for Applied Psychophysiology and Biofeedback* (AAPB), and one of the many regional societies that often hold annual conferences addressing issues related to neurofeedback and biofeedback. We also encourage you to become Board Certified in Neurofeedback. There are a few organizations that provide this certification, however, we strongly support and recommend the *Biofeedback Certification International Alliance* (BCIA).

Here we will discuss information and pointers to assist you in getting started in neurofeedback, as well as some of the pitfalls to avoid. There are different pieces of equipment, different styles of neurofeedback, and a variety of opinions on what works best for any particular condition or symptom you may elect to work with. Therefore, when possible, we will direct you to scientific articles, publications, and research to support information provided here.

Marketing is a cornerstone of any successful business. Whether you're adding neurofeedback to an existing business or starting a new business, this book will provide some basic marketing ideas and suggestions to help you ensure prospective clients understand neurofeedback and how it can benefit them. It will also help you elicit referrals from medical professionals, past clients, and other networks.

We have attempted to keep this book brief but comprehensive. In cases where other detailed written information exists, we will provide you with resources to learn about a particular topic in detail. Otherwise, we do our best to provide you with sound and precise information.

Chapter One

Introducing Neurofeedback into Your Current or New Practice

Why add Neurofeedback to Your Practice?

Years ago, a 19-year-old boy at the urging of his parents and grandmother started neurofeedback training. During one session's set-up process, he casually mentioned that a high school teacher told him he should drop out of school and quit wasting everyone's time. He felt the impact of those words and the evidence of his poor grades, which meant he lived in his parent's home with no plans to take next steps with his life. He believed and had evidence that he was a failure. As he moved through the neurofeedback training process, he began to feel a difference in his mood, motivation, and focus. His father encouraged him to take a drafting course, which would enable him to apply for a job at the company where his father worked. To take that class, he entered a local community college and took several placement tests. He told us with intense satisfaction that he passed two of the three tests and told himself that he could pass that last test. He went home, studied for hours, and then carefully worked his way through the exam and passed. The change in that boy's self-esteem and the dawning understanding that he could work hard and succeed was stunning.

Fast forward five years, when we received a call from his grandmother. She wanted to let us know that he had finished his BS degree, was working at a well-paying job in his field of study and had just started a master's program. This is why adding neurofeedback to your practice is worth it. When an individual's brain begins to work more effectively, they are able to do the work and be successful in the life pursuits they desire. This often results in better relationships, improved job opportunities, and higher quality of life.

Neurofeedback is an evidence-based intervention for many disorders and a recommended 1st-order intervention for ADHD by the American Academy of Pediatrics (Hirshberg et al., 2005). Lawrence Hirshberg (2005) said, "EEG Biofeedback meets the (…) criteria for clinical guidelines for treatment of ADHD, seizure disorders, anxiety, depression, reading disabilities, and addictive disorders. This suggests that EEG biofeedback (aka neurofeedback) should always be considered as an intervention for these disorders by the clinician" (p. 12). Tufts Medical Center researchers conducted a study on 104 school children with ADHD evaluating whether 40 sessions of neurofeedback or cognitive training would improve their symptoms. The neurofeedback group made greater improvements than the cognitive or control group and maintained them six months later (Steiner et al., 2014). Coben et al.(2015) showed that in a study on the impact of coherence neurofeedback training on reading delays in learning disabled children, the experimental neurofeedback group enhanced their reading scores 1.2 grades where the resource room led to no improvement.

It can be a suitable intervention replacing medication or more invasive treatments, and clients under the direction of their prescribing provider are often able to decrease or eliminate medications. Lower medication doses or fewer medications result in fewer negative drug side effects. The effects appear to last, unlike many other treatment approaches that drop off when the training ends or medication is removed (Coben et al., 2011).

It is helpful with disorders after the medical field has done all they can, such as with TBI or where traditional treatment has limited efficacy, such as with ASD. Finally, research has shown that given sufficient training, the positive effects not only remain when training stops but continue to improve over time (Hirshberg et al., 2005).

Conditions helped with neurofeedback include but are not limited to ADHD, ASD, PTSD, addictions, epilepsy/seizures, anxiety/OCD, TBI, insomnia, dyslexia, chronic fatigue, depression, developmental delays, learning disabilities, peak performance, sleep dysregulation, and more. For further information, see ISNR (International Society for Neuroregulation & Research, 2020).

Neurofeedback is a powerful tool. It can be especially valuable when clients have exhausted other traditional treatments without positive effect. As neurofeedback teaches the client's brain to function more effectively, other therapies are often augmented for additional benefit.

Once you have decided to add neurofeedback to your practice, it is important to find reputable training courses and mentors. The following section offers a number of different sources you can access to find what works best for you. This is an incredibly complex and rewarding field, but if used incorrectly can do harm to your client. We recommend that you do not skimp on training.

Training: How do you Choose a Reputable Course

When you consider that neurofeedback is a tool that can literally change an individual's brain, we hope it is clear why rigorous training and mentoring are critical. This tool can do great good and has the potential to cause harm if not used appropriately.

There are many vendors and manufacturers providing a variety of training and courses to those wanting to learn about neurofeedback. In addition, there are associations and a few university courses that offer training.

When choosing a course, we recommend you find one that covers the required Biofeedback Certification International Alliance (BCIA) blueprint for neurofeedback board certification. Whether you choose to become board certified is up to you, but by ensuring you follow their blueprint or take training from an approved BCIA course, you then ensure your chosen course is guaranteed to cover all the critical basics. Board Certification Blueprint information and approved courses can be found at BCIA (Biofeedback Certification International Alliance, 2020).

You have the option to take training classes in person or remotely online. In-person training offers hands-on guidance and direction, as well as the ability to have questions answered in real-time. Remote online classes allow you to work at your own pace and save the expenses related to geographical/travel and cost concerns.

You are combining an understanding of brain function, brain dysregulation, equipment use, and appropriate protocols to teach a person's brain to change its functions. When done correctly, an individual's life can improve dramatically, but as with any tool it must be used correctly to have a positive impact. The research and science regarding the brain and brain functions is rapidly developing and evolving, which in turn provides neurofeedback practitioners with new and improved methods to use neurofeedback. Education, training, and mentoring are the best ways to use this tool for the maximum positive result.

Mentoring: An Ongoing Opportunity for Education

Training courses are a great way to begin learning the basics of neurofeedback. They are designed to cover a core set of information, and then most clinicians find a mentor to continue with the next phase of training. Mentors offer more specific and personal learning opportunities.

Mentoring is critical for applying and extending this basic training knowledge to your practice. Mentors can teach you how to take general information and hone in on the specifics of the population you want to train. They often have access to articles, books, and publications relevant to your chosen client population. They are invaluable when unexpected computer or client situations arise. A mentor can also help you create treatment plans and expertly guide protocol decisions.

As you progress and increase your knowledge base and expertise, you may work with several different mentors. There are a wide variety of neurofeedback approaches, and we recommend mentoring with different experts. Mentoring generally continues until you are comfortable making your own protocol decisions, and then it's about staying current with the latest advancements in the field.

You can find clinicians who mentor at both the BCIA and ISNR websites (BCIA.org and ISNR.org). Even if a clinician doesn't expressly state that they offer mentoring, it may be worth contacting them if they have experience with your clientele and your chosen neurofeedback approach. If you want BCIA certification with a particular mentor, they can send in paperwork to BCIA to be designated as a qualified BCIA mentor.

The two most common mentoring approaches are in-person and remote instruction. Many clinicians utilize both at different times, depending on their particular needs and what is available. There are pros and cons to both that we will discuss in the next section.

In-Person Mentoring

Mentoring can occur in various formats. However, the two most common methods are in person and remote. In-person mentoring offers opportunities to observe logistics of running a clinic as well as practice manipulating equipment. A mentor can deliver feedback and correction more quickly and accurately, adding to the learning experience.

It is not uncommon to fly out and visit a clinic for a few days for an intensive mentoring experience. If a preferred mentor lives close by then, there is the added convenience of in-person interaction without the cost of travel. However, when one is working with a particular piece of equipment or wants to learn a specialized method or form of neurofeedback, in-person meetings are often not feasible and can become expensive given travel-related expenses. In this case, remote mentoring is a good option.

Remote Mentoring

Remote mentoring can be done through a variety of methods such as phone calls, online meeting software such as Go-To-Meeting®, Zoom®, Google Meeting, and similar platforms. Online platforms allow both visual and verbal interaction in addition to the ability to share client data such as EEG information. Client information can be exchanged with HIPAA secure electronic vehicles such as Dropbox and other cloud storage options. The value of remote mentoring is enormous when you consider how quickly one can access specific expertise across distance and time zones without the cost and time of travel.

Mentoring is an important part of the learning process in the neurofeedback field. Both in-person and remote mentoring have strengths and weaknesses. The best option for you will depend on your particular circumstance.

How to Choose a Mentor

Choosing a mentor is important, and several factors should be considered. A mentor who is familiar with the equipment you use, experienced in doing QEEG brain map interpretation, and familiar with the type of clients you want to work with are important factors to consider. A good mentor can help you select protocols, develop protocols, and assist in helping you understand which protocols may work best and why. For more details about selecting a mentor and the mentoring process see, Longo, R.E. & Soutar, R. (2019). Becoming Certified in Neurofeedback. Greenville, SC: FNNR.

Webinars

With the ease of electronic communication, more organizations and vendors are offering webinars. This opens up a unique opportunity for the clinician to pay and attend specific educational lectures without the high cost of travel or time away from their business. There is also the added benefit that many webinars offer a recording that can be watched again to review areas of particular interest. ISNR, AAPB, and BCIA all provide regularly scheduled webinars, and the majority provide CEU credits.

Conferences

Conferences bring together the top researchers, clinicians, and leaders in the field to present their latest findings. Not only do these meetings offer insight into field advances, but opportunities to network and make connections with other clinicians. These meetings also offer the continuing education credits required for licensure and BCIA certification. There are several regional, national, and international conferences provided by the major organizations addressing biofeedback and neurofeedback within our field.

Regional

Regional conferences offer an opportunity to get to know clinicians that practice in your area. These are wonderful referral sources as invariably a client will call that lives too far away from a particular clinic. If the clinician knows other practitioners in their region, they are more likely to refer to a particular clinician rather than sending the client to a general provider list such as ISNR. In addition, getting to know clinicians in your area offers an opportunity to share knowledge and increase the level of professionalism of the field as a whole. For the most part, neurofeedback clinicians have been highly supportive and generous with their knowledge rather than competitive.

Within the USA, several regional organizations provide annual conferences with a focus on neurofeedback and biofeedback. Some of these organizations include the Northeast Regional Biofeedback Society; the Southeast Biofeedback and Clinical Neuroscience Association, the Biofeedback Society of Florida, the Western Association of Biofeedback and Neuroscience, the Hawaii Biofeedback Association, the Mid-Atlantic Society for Biofeedback and Behavioral Medicine, the Biofeedback Society of Texas, and the Western Association for Biofeedback and Neuroscience. Local and regional vendors often have information booths at these conferences.

National and International

National and international conferences offer an opportunity to learn from a wider range of experts. These conferences generally have the funds, reputation, and larger attendee numbers, which generally attract bigger named experts. Attending these meetings offers a clinician the opportunity to interact with many top individuals without the cost of going to each of their clinics. This is a wider breadth of information than what you can obtain from a single training course or one mentor. Hearing these experts present and meeting individuals in a more casual setting can offer you the information you need to find your next mentor and add new advancements to your practice. National, regional and local vendors often have information booths at these conferences.

Annual national and international conferences are offered by The International Society for

Neuroregulation research (ISNR); The Association for Applied Physiology and Biofeedback (AAPB); and The Biofeedback Federation of Europe (BFE).

Board Certification

A clinician who becomes board-certified in neurofeedback and/or QEEG offers the public or other medical providers a way to measure credibility, validation of skills and knowledge, and adherence to ethical standards. It includes coursework, mentoring hours, and an exam. BCIA certification is now written into an increasing number of state laws.

Biofeedback Certification International Alliance (BCIA)

We believe that anyone practicing neurofeedback should pursue board certification through the *Biofeedback Certification International Alliance* (BCIA). BCIA is the only internationally recognized certification body for the practice of biofeedback by the Association of Applied Psychophysiology and Biofeedback (AAPB), the Biofeedback Federation of Europe (BFE), and the International Society for Neuroregulation and Research (ISNR). Some vendors provide certification through the use of their particular product, but they often do not provide the level of learning and requirements required to become certified through BCIA.

For further information about the BCIA mentoring process see, *Becoming Certified in Neurofeedback* (Longo, R. E. & Soutar, R., 2019).

International QEEG Certification Board (IQCB)

Over the past few years, certification in QEEG has been developed and is becoming more popular. This certification is important when the practitioner conducts a large number of QEEG brain maps and conducts such for other practitioners who do not engage in QEEG brain mapping. At this time, certification is offered by The International QEEG Certification Board (IQCB).

Chapter Two

Your Brand and Marketing Avenues

Your Brand & Value Proposition

What is your brand? Your brand is what your potential customer thinks of when he or she hears your brand name. It is the collection of feelings, emotions—the intangible sum of all the attributes of your services. Your value proposition is a declaration of what you stand for, what value you provide, and why clients should do business with you (perhaps instead of a competitor). So often at neurofeedback conferences, speakers and attendees talk about the impact this treatment has on their clients and the profound impact on their client's families, social circles, and future prosperity. As clinicians work to communicate this potential, they are in essence trying to define their brand value proposition. A great value proposition describes succinctly what you do and the top reason(s) why a client should avail him/herself of your services. For example, one clinic's brand promise is that we help you live life as it was meant to be (or excel in school, work, or socially) as we train down or eliminate the symptoms of ADHD, ASD, depression, and more, without the use of medication.

Brand identity is the face of your brand. It is what the customers can see such as your logo, key colors, typography, photography, and the tone of your written work. It is important to have a consistent brand identity across all parts of your business. What does your physical office look like? Some have chosen a more traditional office space modeled after a medical office or hospital. Others have created a warmer, more intimate physical office space. Do you call your customers "patients" with the connotation of someone who is ill or "clients" with the connotation of a customer accessing a service? Is your website easy to use and up to date with the current methods of communication across computer, tablet, and phone? What color scheme have you chosen, and are you aware of the emotional impact of different types of colors? The critical piece is that you consciously choose how you want your customer to perceive your business, how you present your business and ensure your brand identity is consistent across all your marketing materials and avenues. We'll discuss a number of these issues as we continue through this book.

Branding & Marketing Avenues

Branding is the process of building awareness, loyalty and engagement, which brings clients to your door and awareness to your community. It allows your business to stand out and be distinct from other similar businesses or from other alternative choices. It involves marketing, advertising, and selling via many different avenues.

Marketing your neurofeedback services is extremely important. There are many different avenues you can use to market your services. Neurofeedback is not a common service, like roofing, plumbing, or electrical work, where you can often merely advertise the name of your service, and people immediately understand what you do. Neurofeedback marketing often requires a strong educational component. This excludes some marketing avenues that rely upon immediate category recognition and do not offer space for educational content, such as banner ads, school program ads, etc. We will describe some of the most valuable marketing avenues that allow an educational component.

Before you dive into branding and marketing, you need to consider the audience you'll serve, the disorders you'll specialize in, and the scope of your practice based upon your degree and licensure.

Target Audience

Who is in your community or within driving distance? Working professionals? Mothers of young children? Parents of teens? Grandparents? For example, one practice in a large city is well known for nationally ranked elementary and secondary schools. The community makeup is very heavily weighted towards families, enabling them to focus heavily on children and, to a lesser extent, on younger mothers. The practice is in a bedroom community outside of a large city but has not chosen to focus on technology or other working professionals in part because the commutes are so long, preventing them from attending sessions except for the very late evening. If you're not already intimately familiar with your potential customers' demographic makeup, speak with your Chamber of Commerce or your local Service Corps of Retired Executives (SCORE) to help refine our target audience.

Disorders

In general, neurofeedback can be used to address disorders of the brain and central nervous system. Some practitioners work broadly with a variety of disorders without a specific specialization. Other neurofeedback practitioners narrow their focus to train specific disorders. For example, one clinic began focusing on children with ADHD, ASD, and some other disorders. They fortuitously made inroads into a very large community of those with ASD. They've also made a name for themselves amongst therapists and counselors in public and private schools after they demonstrated remarkable turnarounds with their most problematic children with ADHD, which now make up a large proportion of their clients.

Scope of Practice

Scope of practice is one of the areas in which a practitioner can easily get him or herself into trouble.

Be sure that the disorder you are working on is within your scope of practice or work with a specialist who can provide you with clinical guidance or supervision for a particular disorder. Neurofeedback clinicians may elect to address the symptoms of anxiety, depression, relaxation, and sleep problems. Therapists who have a particular specialty such as addictions may focus their neurofeedback training on this particular group of clients.

If you plan to limit your practice to specific disorders, then it is best to note your specialty areas in your marketing materials.

We encourage practitioners to use the term "brain wave training" versus "neurofeedback treatment" to avoid stepping outside your scope of practice. "Training down negative symptoms" is one way to phrase the use of neurofeedback versus saying that one is "treating a patient for ADHD," which is a diagnosable medical disorder. As seasoned neurofeedback clinicians, we have seen some of our colleagues have their practices closed because they were "treating" a medical disorder and thus charged with practicing medicine without a license.

Website

Potential clients typically search and gather a great deal of information online. Your website is imperative, as it is often their first impression of you and your services. Adweek.com estimates customers are 81% of the way through the decision-making process before they speak with a real person on the phone or in-person (Morrison, 2014). Buyers in all industries are using search engines, social media, and website content to make decisions. They are looking for information that educates and informs and helps them believe you can solve their problem.

Your first step is to choose a domain name (typically as close to your business name as possible) from

GoDaddy, or another domain name registrar. Try to avoid names that are a dash or period different from a competitor, as this may inadvertently drive business away from your site. Neurofeedback-Center would be a poor choice because people will forget the dash and may end up at your competitor's site.

Before searching for an individual or company to help you build your website, think through what you want from your site and your preferred platform. Currently, WordPress is an extremely widely-used and highly regarded platform as it is easy, adaptable, expandable, and relatively low cost. It has a huge number of templates and themes that you can choose from that allow you to quickly build a website with minimal effort. You can make changes to or add new services, graphics, text, and blog posts with a couple clicks of the mouse as opposed to needing a graphic designer or web designer to make changes to your website. WIX, is another similar program.

You'll need a web hosting company such as SiteGround, InMotion, Register, Bluehost, or GoDaddy, which are all highly rated, reasonably priced (typically $7-10/month), and have great customer support. Most web hosting companies offer a one-click WordPress installation, and many of them have tutorials on how to configure and set-up.

You'll need to decide how complex or simple you would like your site to be and obtain bids from several web designers to find a fit that works for your goals and budget. As with any service, researching good quality companies to build or upgrade your website is important. The process of obtaining bids from several web designers can help you understand what is possible and what the average rates are for different size projects. Look for designers who can create a cohesive brand identity, user-friendly design, have strong experience with WordPress, and are highly rated or recommended.

Anywhere from 30-70% of web traffic is now from

mobile devices. As well, Google penalizes websites in their search ranking that are not designed for mobile use. Thus, your website should use responsive design principles, so it is accessible and attractive on the computer, tablet, and phone. Most newer WordPress themes enable responsive design principles.

At the barest minimum your site should include a welcome or home page, information about each of your services, directions, testimonials, resources (i.e. where to learn more about neurofeedback, research, etc.), contact information, and an about us page introducing you and your staff.

Clients and peers have strong preferences about acquiring information through electronic or paper form. It is valuable to have both available to reach each audience and know their preferences.

Brochures and Business Cards

School, community, and business networking events are venues where paper business cards and brochures are expected and valued. Parents who have children struggling with ADHD, ASD, or learning disorders may take a brochure home and with this physical reminder look up more on your website and call you. Business card exchanges are the norm at business networking events and offer an easy place to jot a note down about the person or follow-up meeting details.

When putting together a business card and brochure, local graphic designers and printers are excellent resources. They may give you more for your money with personal attention and design assistance. On the other hand, large online printers such as Vistaprint have a wide selection of templates and are known for inexpensive printing. They also offer a variety of marketing products and can easily imprint your business logo and information.

When creating a brochure, consider that tri-folds are easy to put together and many templates are available in Microsoft Word. While this may fit the needs of your business, it is used by a wide range of people such as the neighborhood dog walker, teenage window washer, or the neighborhood bake sale. Booklets (several pages cut and stapled in the middle) are considered more professional and offer you another way to stand out. Booklets can be sized so that they fit in an upright plastic stand easily accessible in your office waiting room or at a public event. This size can also fit in an envelope and can then be used for direct mail efforts with the addition of a professional letter on business letterhead. Your business cards and brochures should be consistent with your website and the rest of your brand identity.

Business cards and brochures are often used at school, community, and business networking events. Both local graphic designers and printers, as well as online printers offer a variety of options for you to evaluate. Remember to keep the design of these paper marketing materials consistent with the rest of your brand identity.

Social Media Networks

Social media can be used to build your brand, build awareness, connect with prospective customers, educate prospective customers about your services, and create community. Different audiences gravitate to one form of social media over others, making it very important to clearly understand what segment of the population you are trying to reach so you can post your information where they are commonly found online. Facebook is overwhelmingly the most widely used platform in the US, and users tend to skew middle-aged to older, urban, college-educated and slightly more affluent people. https://sproutsocial.com/insights/new-social-media-demographics/ Instagram tends to skew younger, slightly more rural, with fewer college educated. Twitter users tend to be slightly older, urban, more college-educated and relatively affluent. Each social media type has different "social rules" that govern acceptable length and type of information sharing. Be sure to learn about these rules so that you don't inadvertently break social norms and alienate potential clients.

Facebook can be a great place to reach a large portion of the population in regional communities made up of your target clientele. The online community pages are typically where people gather to get information about their local community they live in, including recommendations for doctors, dentists, mechanics, and other service providers, as well as to keep informed of current events or warnings of crime incidents. In our suburban area, there are typically two and sometimes three Facebook pages serving each community. Preference should obviously be given to groups with larger membership. One group has nearly 20k members, one has seven thousand but low traffic, and the third, although only having four thousand members is very high quality and carefully moderated, making it very valuable to us. In our experience, clients have generally been willing to drive a maximum of 30 and sometimes 40 minutes to our clinic for biweekly neurofeedback appointments. Knowing this mileage limit, the Facebook communities within an approximately 20-minute drive are strong candidates for marketing.

If you haven't already created a personal Facebook profile with your name, photo, and details. As well, you need to create a Facebook page for your business. You can limit privacy settings for your personal profile as you see fit, but your business page should be very public to enable people to find you easily.

With your personal Facebook account, find and read the posted rules in your local Facebook community groups. Some do not allow advertising, others allow advertising only on certain days, and others might allow advertising only once every week. Join the larger or more valuable groups and start interacting with the groups to establish a personal presence.

You can participate by helping others who are asking for resources or advice as well as looking for information yourself. If someone posts looking for services to address a problem that you offer, such as kids struggling with ADHD, you can respond with advice or a recommendation for your business. The goal is to be an active community member while making sure you have made enough posts to get rid of the new member flag before making your first business post, typically in about two weeks. If you only advertise your business without additional community participation, it is generally looked on as spam and reflects poorly on your business.

If the rules allow, craft a post describing what you do for your desired audience and invite people to call for a free consultation. Encourage current or past clients who're also in the group to tell people what they have experienced at your office, tagging your business name in their post. You can post a testimonial and invite those that would like to see similar results to contact you.

Facebook wants a business to use a business account rather than a personal account. There are some advantages, such as tracking features that can help you understand the demographics of your audience and other analytics. You can also run ads with specific demographics. However, Facebook suppresses business posts in favor of personal friend posts. What that means is that you typically need to post tips and helpful information on your business page and share those posts from your personal account. This social media vehicle moves rapidly, and Facebook rates businesses on how fast they respond to comments or requests. If you cannot monitor and respond quickly to posts, consider asking another employee to handle this responsibility.

There are similar best practices for Instagram, Twitter, and other social networks. Social media outreach is very valuable and a necessary part of marketing your business. It offers you a place to reach your target audience at very little cost, but you must understand the unique social rules of these platforms to navigate them well.

Testimonials

Testimonials are critical for the long-term growth of your business. People trust others just like themselves far easier than anyone selling goods and services. Testimonials are written or recorded statements that support your credibility, expertise, and build your reputation. Testimonials offer not only information but through stories, they generate emotional appeal that can build trust and drive action on the part of a potential client. For example, Amazon reviews are incredibly successful because customers can see the experience that other consumers have had with a particular product.

When asking for or selecting a testimonial, remember that detail and storytelling are compelling. A client who says, "They were great!" is helpful, but a more compelling testimonial would say, "My son struggled with ADHD and was failing out of grade school until we sought help from X company and watched neurofeedback turn his life around. Now he brings home straight A's and loves going to school." The best testimonials describe the before and after neurofeedback state in a way that helps people relate to the experience and desire the same outcome.

Make sure you are getting testimonials from your target audiences. If you work with seniors, make sure they mention their age or age range. If you are working with children, make sure their parents talk about the impact on both their children and their own lives. Testimonials can be especially powerful if they are comfortable letting you use a picture. Better still, if you can get someone to record a brief video describing their experience, the testimonial can be extremely powerful on your website or social media. Testimonials should be repurposed in in all other marketing vehicles such as (a) website, (b) Facebook/social media, (c) brochures and business cards, (d) networking groups, (e) references and referrals, (f) blog, (g) promotional items, (h) presentations, (i) summary reports to providers.

A great time to ask for a testimonial is when you hear a parent talk about telling their neighbors or their child's teacher or all their extended family about neurofeedback. If they are openly talking about it to others, there is a high likelihood that they would be comfortable giving you a testimonial. Another indication is if they talk about wishing other people knew how much neurofeedback can do for them. You can verbally ask or have a written request as part of your paperwork. It is important that you make it clear they are not obligated, and there are no negative repercussions if they decline.

Be aware of and ensure compliance with HIPAA regulations. It's important to obtain verbal and written consent to share testimonials. You can list their name or just their initials. You can include a picture, or you can use a stock photo for attention. Some may be willing to have their testimonial used in your brochure with limited exposure; others may allow their testimonial to be used on Facebook. There are a wide variety of ways to share information, and you can gauge who is comfortable doing so and who prefers to remain more private.

Networking Groups

The goal of a professional association is to share

information and connect members. By joining a networking group, you can receive referrals who are potential clients or others who can offer opportunities such as speaking engagements, publishing opportunities, or referral sources. There are a variety of different types of groups, and it is essential to examine what best fits the needs of your business. Consider when choosing a group: time commitment for participation, cost of membership (both yearly fees and individual activities), group membership, exclusivity (i.e., multiple members with similar businesses, members sharing a similar licensure, or industry exclusive), the stated goal of the group (community service, business building etc.), and the opportunities for personal growth and development.

Regular attendance helps to form and strengthen relationships that encourage the sharing of business opportunities. When looking for a networking group, visit as many as possible within your area to get a feel of their goals, culture, and membership. Neurofeedback is not as well-known as financial planning, dentistry, or cosmetology. Consequently, it takes time to educate people about what neurofeedback can do for members and their connections. Obtaining referrals and recommendations from general-purpose networking groups may require some patience and commitment.

Ivan Misner and Brian Hilliard break networking groups into five main categories (a) casual contact networks, (b) strong contact networks, (c) community service clubs, (d) professional associations, and (e) online/social media networks (Misner, 2018).

Casual Contact Networks

These types of networks allow business groups, both large and small, and multiple members of the same kind of business to join. They usually meet monthly with an opportunity to mingle informally and then have guest speakers who talk about business, legislative or community topics. The Chamber of Commerce is an excellent example of this type of group.

Chamber of Commerce groups typically charge a price for membership, which includes a listing on their website and then additional fees when you attend their monthly luncheon or other events. Their meetings usually offer the opportunity to network as the meeting starts or during activities, but also have a variety of speakers or other planned activities. They are not industry exclusive, and businesses can join several chambers in different towns. These groups range in usefulness depending on your networking skills and the inclusiveness of members.

These groups are not primarily organized around referral generation, which means each member needs to hone their skills at networking to accomplish this goal. There are many resources available such as the book, *How to Work a Room*, which can teach the skills of efficient and effective open networking. Another way to maximize exposure in this type of venue is to volunteer to be a chamber ambassador or join one of their committees. Volunteering allows you to help the organization while developing relationships, which could turn into business opportunities in the future.

Strong Contact Networks

These are businesses whose primary purpose is to help members exchange referrals. They typically meet weekly, early in the morning or over lunch. Membership is typically restricted to one member per profession.

Business Network International (BNI) is an example of this type of group. BNI or similarly-styled groups have several key characteristics. They are industry exclusive, which means only one lawyer, one electrician, or one neurofeedback clinician is allowed to join the group. Though you may not have a great deal of competition with other neurofeedback providers, it does ensure a wide range of other professions who then have a diverse group of contacts that may be beneficial to your business. Continued membership is dependent upon a very high attendance rate as well as the number of referrals passed to group members.

At each meeting, the members stand up and give a ~45-second commercial to educate and promote their business, and to ask for specific referrals. A rotating schedule of 10-minute presentations allows members to share more detail about their business and solicit referrals. The group encourages periodic one-to-one meetings outside the regular group meetings to help build relationships and understanding of what each company offers.

There are many positive benefits to this networking model. You have the opportunity to educate others about neurofeedback and what it does, build relationships, and encourage referrals. The weekly short commercials give members practice stating what they offer in a concise and interesting manner. Consider the cost of membership, how many businesses are members (a group of 25-40 is optimal), and time commitment when assessing the value of this style of a group.

Community Service Clubs

Service groups such as Rotary Clubs are not focused on referrals but instead bring together business and professional leaders to provide service in the community and the world. There are costs to join and additional fees for various meetings as well as time spent volunteering. This organization benefits the community with service and can be an excellent way to give back.

During the different service activities, members have the opportunity to form relationships with other businesses and promote positive business name recognition in the community. Membership in a service club is a less direct way of increasing business. Consider the cost and time spent volunteering when evaluating the most efficient way to grow your business.

Professional Associations

Professional associations typically include members within a single industry. ISNR is such an example with the stated goal of exchanging information and best practices amongst neurofeedback and biofeedback professionals. ISNR has the largest membership of any professional organization in the neurofeedback field and presents a wide range of topics and speakers at their conferences. A great way to maximize membership in these groups is to attend conferences, participate in forums, and serve in leadership positions.

Professional associations can also be less formal, small, and local. One such group in my area is a group of various practitioners who informally meet at an ASD clinic. This group has less than ten members, including the clinic owner, therapists, coaches, and neurofeedback clinicians. We discuss various issues, share resources, and refer to each other as appropriate. This group has not only helped with referrals and business contacts but has become a social and emotionally supportive environment. There is no cost, and we meet on a monthly basis with occasional emails during the month.

There may be other networking groups or opportunities that are unique to your area. The primary consideration is whether an organization allows you to accomplish your business and personal goals in the most time and cost-efficient manner.

Educational Speaking Opportunities

Educational presentations at local organizations, schools, health fairs, and hospitals or clinics are an effective way of both educating the public about neurofeedback and gaining clients. The degree to which you educate vs. sell during the presentation, the presentation length and type can vary widely depending on the group, the time available, and the knowledge level of the audience.

Many organizations will prefer that you add value to their audience by educating them by providing information they can use to improve their lives. For example, you might talk about six ways in which you can help the ADHD child remain focused on homework or helpful hints on structuring the environment to calm the ASD child. You can talk about how the brain works: how the ADHD brain responds to stimuli or how the ASD brain misses cues. And of course, you can talk about neurofeedback and how it might provide an alternative to some therapies. The level of promotion of your services should be tailored to the desires of the event organizer.

Many networking groups offer opportunities to deliver a five to ten-minute presentation on a regular basis. Local organizations, schools, and community fairs have opportunities for local businesses to present to their audience when the topic is of interest to their group. Typically, the presentation in these groups is 30-45-minutes with time for questions. Hospitals and clinics also periodically offer opportunities in their health and wellness fairs for you to speak briefly as well as operate an information booth.

Your presentations should be polished and professional. If you use a PowerPoint presentation, work with your designer to ensure your slide template complements your brand image you've established with your website, business cards, and brochures. Make sure the slides convey a concept without being overly cluttered with text. You want the audience to focus on you and your message rather than reading along with you on the slide.

Speaking opportunities are the perfect time to offer

promotional items such as pens, stationery, or other items with your company name and website imprinted. Even more importantly, you should obtain attendees' names and email addresses to add to your mailing list and other marketing campaigns. You may use a sign-up sheet for those that would like more information. Or you might consider offering a give-away in exchange for their email address. Some practices have used to great effect a brain-shaped stress ball that the kids love. They have offered to send them via email a series of ten tips to tame the ADHD brain. Many practitioners offer a free consultation in exchange for their contact information.

Referrals

It is important to ask for referrals from your network of friends, providers, associates, and especially clients. Typically, these referred individuals are, to some extent, pre-sold. Rather than needing the seven-plus contacts with the idea of neurofeedback, they are ready to sign up with one phone and perhaps a visit to your office.

When participating in a networking group, part of your job is to ask for specific types of referrals. When you give them specific and descriptive examples, they are more capable of knowing who you can help and referring or introducing them to someone in need. For instance, "I'm looking for parents with ASD children who are struggling because they have no friends." Another example, "Do you know a parent whose child can't sit still or focus for two seconds? Who can't manage to turn in their homework even when it is done because they forgot to put it in their backpack?" At the same time, you will be listening and doing the same type of thing for them if you respect them and their product.

Local providers are more likely to refer to you if they are educated about what neurofeedback can do and how it compliments their business. As you build relationships through networking, community events, and other interactions, local providers can see how combining neurofeedback with what they offer a client increases overall improvement and healing.

A great time to ask for a referral from a client is after they give you a testimonial because they're so excited about the progress they can see. If you have improved their or their child's symptoms and can also show measurable QEEG changes, people are understandably excited. This can flow naturally into an offer to help others they know who could benefit.

Promotional Items

Promotional items are a great way to encourage a customer to talk to you and walk away with an item that will remind them of your business and what you can do for them. A well-done item is sticky (meaning that people keep it, use it, or play with it) for a long time. Cost can vary, and companies are often willing to price match or honor a previous price for return customers.

Make sure you know your target audience and what they would find interesting. When and where are the customers going to receive the promotional item? We used brightly colored brain balls at a school health and fitness fair. The promotional items are visually interesting and attract the kids, which pulls the parents to our booth, where we are able to start a discussion.

Consider how commonly used some items are to ensure you don't get lost in the noise. Pens are quite common and often cheap. However, a local home inspector used a higher quality pen with a flashlight embedded in the cap. That pen has stayed in our office and frequently reminds me of this man. The following are a few of the more common or popular promotional items (a) writing instruments, (b) tote bags, (c) journals and sticky notes, (d) lunch bags, (e) stress balls, (f) hats, (g) sunglasses, and (h) drinkware. When ordering, make sure you are aware of how much time the company needs to produce and ship the promotional items to you.

Promotional materials that have stickiness keep working for you long after you might imagine. A practitioner we know reports she had a man wander up to her in a town event and show her his stress brain ball that he had had for several years. At a local school fair, the second year she attended, she had two people tell her they had her information from the previous year and were finally able to move forward with training. One child told his mom (who was the PTO organizer of the event) that he liked "The Brain Lady table."

When you have organic growth from both children and adults, the clients show up at your door. Do potential customers have to do anything to obtain the item? We ask them to leave an email address, take a photo with the item and post on social media, or leave a review on our website. Make it easy and fast to take an action by using a quick form on an iPad or offering a business card drop container.

Summary Reports to Other Providers

A summary report is both a professional courtesy and another way to let those working in the medical field in your area know that you offer neurofeedback. We ask the client during the intake process to provide the name of their primary care, referring physician if different, therapist as applicable, and offer to send them a copy of the report. Providers gain a better understanding of the neurofeedback process and measurable improvement, which in turn helps their modalities have a more significant impact. Parents find this information helps them advocate for their child within the school setting and IEP meetings. The report includes intake information and QEEG results. We send out a report at the beginning of training and every fifteen sessions after that with a progress summary and highlights from the most recent QEEG results.

Whether you are adding neurofeedback to an existing practice or starting a new neurofeedback practice, your brand and marketing avenues are critical to spreading the word that you offer a valuable service. As a reminder, stay within your scope of practice, ensure you are HIPAA compliant, and beware of FDA guidelines on "treatment" versus "training."

Chapter Three

The Initial Meeting or Consult

Initial Contact

A potential client may make initial contact through email, phone, Facebook, or in person. Some may have specific questions that you can answer, or they need an overview and a more in-depth explanation of the process. In many cases, there are questions that come up routinely. You should write out several scripts in advance to respond to common customer inquiries. You can refine these as the need arises. You can copy and paste these in response to email inquiries from potential clients, and you can place them on your website to answer people's questions before they contact you directly. Here is an example of one such script for a general inquiry.

Name,

Thank you for reaching out to ask about neurofeedback. When someone comes to see us, we do a verbal intake assessment, a short computer test, and a QEEG or brain map. The QEEG data is compared to a database so that we can see how the client's brain is functioning compared to the norm in his or her age category. This allows us to focus the neurofeedback training in specific areas that need the greatest help. The actual neurofeedback training sessions are twice a week on either Monday and Wednesday or Tuesday and Thursday. We do fifteen sessions and then conduct another brain map. This tells us how far things have shifted and where we should focus next. The average client typically does two to three sets of fifteen sessions.

We are not affiliated with any insurance company though we always encourage our clients to see if they can receive reimbursement. Our receipts list our tax ID number and a CPT code for biofeedback (neurofeedback is a type of biofeedback). We accept cash, check, Visa, or MasterCard.

We charge $_ for the brain map, $_ per session, and offer a discount of $_ if you pay in full in advance.

I would be happy to speak with you further, or set up an intake assessment appointment if you would like.

Warmest regards,

The First Meeting or Consult

When providing neurofeedback services to clients, we encourage you to have an initial meeting or consultation with the prospective client. Many providers offer a free consultation. During the initial meeting or consultation, the clinician explores how the prospective client learned about neurofeedback, the reason(s) for wanting neurofeedback, how the person came to that decision, and whether they are a good fit for your clinical focus. This is important information to understand because many clients have a disorder for which they have tried medicines or other interventions that have not worked. Some come to the office considering neurofeedback as the last resort.

During an initial meeting or consult, there are several items the clinician may want to cover, including:

1. How they learned about neurofeedback.
2. Their reason for wanting neurofeedback.
3. Are they a good fit for your clinic's focus?
4. Taking a brief health history.
5. Discussing what a QEEG brain map is (when the clinician performs this type of assessment).
6. Other diagnostics or testing.
7. An overview of brainwaves and their function.
8. How neurofeedback works.
9. Number of sessions.
10. Length and frequency of sessions.
11. Side effects of neurofeedback.

12. Costs for QEEG and/or neurofeedback sessions.
13. Factors that negatively influence neurofeedback.
14. Lifestyle changes that support neurofeedback.
15. Whether you accept insurance, or the client is responsible for seeking reimbursement.
16. CPT codes and flexible/medical spending accounts.
17. Tour of the facility, equipment, and introduction to staff.
18. Next available openings.

Once you have been properly trained in the basics of neurofeedback and have had several hours of mentoring, the above items will be easy for you to discuss with a potential client. As a part of being mentored in neurofeedback, BCIA requires the mentee to review the *Neurofeedback Essential Skills List* (Biofeedback Certification International Alliance, 2020).

Intake Process

Once an individual decides to move forward with neurofeedback training, a consent must be signed, an in-depth verbal intake completed, and any additional testing or evaluations administered. The consent puts in writing the specific details of the agreement between both parties. The verbal intake incorporates the questions asked during the first meeting or consult but in far more depth and with other questions. Additional testing may include cognitive tests and a QEEG.

Consent and Agreement Forms.

When the prospective client agrees to move forward, an informed consent form should be reviewed and signed by the client. Many clinics have various forms for the client to sign that may include the client's rights and responsibilities, HIPAA compliance, financial agreements, etc. Be clear about consent for provider to provider communication, family member communication, and remember that age of consent may be different from state to state. Risks, benefits, costs, and time commitment for training should also be stated. A consent form not only protects both the provider and the client, but clarifies expectations for both parties.

Payment Terms

Payment terms often include a session by session cost and if offered, a discount for paying in advance for a block of sessions. Be sure to evaluate the terms and fees you incur to decide which credit cards you will accept. Generally, Visa and Mastercard have lower fees than American Express. Also, be informed of specific payment parameters if you use online credit card payment processors such as Square, PayPal, or Stripe. They may place holds on funds if entering credit card numbers above a certain amount versus swiping/chip and signing in person. Find the best option that fits your business needs and periodically re-evaluate as new options and updated offers frequently become available in the business world.

Late/cancellation policies are also important to include in your consent. These can evolve over-time as particular issues arise. For example:

Late Policy

It is important to us that you not be kept waiting, and we do our best to honor your time. Neurofeedback sessions run for (session time) minutes. If you are late for an appointment, we will do our best to fit in your session and give you as much attention as possible through the end of your scheduled appointment time. To be fair to the clients following you, however, your session will end on time. You will be billed for the full session, regardless of the time you arrive.

Cancellations

Our schedule is very busy with clients who desperately need Neurofeedback. If you find that you need to cancel or change an appointment, please give us 24 hours' notice so we can offer Neurofeedback to another client during that time. We will make reasonable efforts to reschedule sessions which are cancelled in a timely manner. If you cancel or change an appointment without 24 hours' notice, you will be charged the full fee for the session you missed.

Pausing Neurofeedback Training

Neurofeedback Training is most effective when it continues uninterrupted. If you choose to pause training, we cannot guarantee your time slot will still be available. If you pause for an extended period of time, we may need to conduct a fresh QEEG so we can tailor training. When you are ready to restart, we can discuss options and availability.

You can always choose to waive a fee or make an exception, but having the policy gives you the option to give appropriate consequences that encourage clients to come consistently and on time for training. This in turn, allows you to efficiently help clients and run a profitable business.

Intake Assessments

Depending on one's setting and specialized office procedures, the prospective trainee will need to fill out forms and health-related information. Many clinicians conduct a verbal intake assessment that typically covers medical, psychological, social, academic, and work history. The goal is to spark the client's memory to elicit key information. They may answer several questions in one long explanation. Understand your intake well enough that you can jot down the information and not be so stuck following your form that you ask questions they have already answered.

Be aware and communicate with the client issues that can decrease the effectiveness of neurofeedback. Multiple psychotropic drugs may decrease progress. Medications that inhibit memory and or learning can have a negative impact because neurofeedback is a learning process. The same applies to recreational drugs. For instance, marijuana produces significant deficits in attention and memory. A client who is considering committing time and money to neurofeedback training should be made aware of this effect and its long half-life (Volkow et al., 2016; Levinthal, 2016, p. 138; addictionBlog. org, 2011). By educating the client, they can speak with their doctor and have their medication adjusted or make personal behavior changes with recreational drugs. It gives them the ability to make an educated decision. Take into consideration significant health issues that may come up during the intake process and refer to the appropriate medical professional when needed, such as for sleep apnea, medication concerns, or seizure activity.

Additional Testing or Evaluations

Cognitive tests offer on task performance and insight into brain function. Clients may have neurological or psychological test results from other medical providers or take tests from neurofeedback clinicians with appropriate licensure. Additionally, the client can take cognitive performance tests online. Two such cognitive tests available at this time are CNS Vital Signs assessment and Cambridge Brain Sciences assessment. They differ in price structure, user interface, full test/ partial test options, and report formats. Some QEEG Brain Mapping systems and software offer similar forms of testing.

Lastly, modern, effective training typically includes a QEEG. The electroencephalogram (EEG) is the measure of brainwaves at the scalp, while a quantitative EEG (QEEG) is the analysis of the digitized EEG. QEEG analyses can show brain wave activity patterns that may align with clinical symptoms (Sherlin, 2016). QEEG results increase the clinician's ability to correctly set up a protocol that addresses the areas and issues the client wants to address. Symptoms can be misleading, and very different QEEG results may come from clients with similar symptoms. By obtaining a solid QEEG in addition to the other intake information, the clinician has the best information to make solid protocol choices. Additional and detailed information on the QEEG is covered in the next chapter.

During the intake process, the information gathered involves a verbal intake with a clear understanding of the client's priorities, cognitive, neurological, or psychological tests, and a QEEG. This three-pronged approach offers a solid way to make excellent protocol choices. The next chapter dives further into the QEEG process.

Chapter Four

The Quantitative
Electroencephalography (QEEG)

Incorporating a QEEG as part of an assessment and protocol selection process allows more specific and measurable data along with other objective and subjective information. A clinician can choose to find a provider to record and process the data, record the EEG themselves and send it out for processing, or record and process the EEG themselves.

If looking for a provider to perform the EEG and analysis, search for one who is board certified or has a high level of expertise. If they are within a reasonable travel distance, the client can go to the clinician's office who offers these services or a trained tech can come to your office to do the recording and then send the data out for processing.

If the clinician wishes to perform the EEG recording and process the data themselves, they should obtain the training and check that this is within their licensure to perform. The International QEEG Certification Board (IQCB) website lists a number of groups who offer training, which also meets a portion of the requirements towards their QEEG board certification. Like BCIA, this includes a healthcare license, didactic coursework, mentorship requirements, and an exam.

We will discuss this section with the assumption that new practitioners are either sending their clients to a more experienced clinician for the QEEG or recording the EEG and sending it out for processing.

For more detailed information, see: *A Consumer's Guide to Understanding QEEG Brain Mapping and Neurofeedback Training*, Longo, R. (2018). FNNR.

In some QEEG mapping systems, much of the health-related information including physiology and metabolic symptoms, cognitive and emotionally related symptoms, and other self-inventory items can be filled out online.

BCIA requirements to become Board Certified include conducting a proper intake assessment, as well as an understanding of the EEG and the QEEG process. Here is a portion of the skills list from BCIA:

Intake, Assessment, and Protocol Selection

1. Document a thorough client symptom and medication history and gather background information relevant to treatment/training goals

2. Provide a thorough EEG baseline assessment, using the following skills:

3. Perform correct measurements to name and locate on the scalp each of the International 10-20 System electrode placement sites

4. Properly prepare scalp and ears and attach electrodes to selected assessment sites or attach an electrode cap if doing a full-cap quantitative EEG

5. Correctly perform all steps to collect a QEEG recording or multi-channel EEG assessment: checking impedances, removing artifact, and collecting eyes-open and eyes-closed data

6. Demonstrate basic understanding of a QEEG assessment report, as well as the most commonly reported components of QEEG databases (absolute power, relative power, phase, coherence, z-score comparisons, etc.)

7. Identify recordings indicating spike and wave activity requiring consultation

with a neurologist or QEEG expert

8. Use all intake, psychometric, and baseline EEG assessment data to select target electrode placement sites and montages for neurofeedback treatment/training

9. Select an initial neurofeedback protocol and explain rationale to client.

Client Orientation to QEEG

Prospective trainees may become anxious regarding having a QEEG. For some, it is the concern that the QEEG will confirm what they suspect may be the problem, while others fear you may find or discover a problem they don't believe exists. In general, most clients look forward to the process and setting up a plan to improve their health.

It is our belief, due to the circadian rhythms of the human body, that QEEGs should be conducted during morning hours. Research has demonstrated that EEG varies during the day and noon and afternoon times are often times when brain waves, especially slow waves are more likely to be elevated. Most people are most awake and alert during the morning hours.

The client should allow a minimum of one hour for the QEEG to be conducted, which includes preparation, recording, and clean-up. They should be encouraged to wear comfortable clothes for the procedure and advised that a small amount of conductive gel used to conduct the QEEG will remain in the hair.

In order to help prospective trainees to prepare for a QEEG, a basic instructional set is given that covers the following.

QEEG Preparation Checklist

Providing instructions to the client to review and follow

before they come in for a QEEG will help assure that the best results possible are acquired.

1. Illness ~ If the client is sick, instruct them to call to reschedule even if he or she only has a cold. Coughing, runny nose, and general malaise will result in poor readings for a QEEG.

2. Sleep ~ The client should be instructed to get a good night's sleep before the QEEG. When the client arrives for the QEEG, check to see how well he or she slept. Many clients pursue neurofeedback due to insomnia and other sleep problems. You may choose to go ahead with the QEEG, but tracking their sleep quality will help you track improvement as neurofeedback training progresses.

3. Hair & Scalp ~ The client's hair needs to be clean and dry with no product in it (i.e., crème rinse, conditioner, moisturizers, mousse, oils, hair sprays, or hair gels). Clients should wash their hair three times with a ph neutral shampoo such as Neutrogena Anti-Residue or Suave Clarifying shampoo before the appointment. Be sure that the client's hair is dry before starting the recording, as damp hair can interfere with the accurate collection of the QEEG.

If the client has hair extentions, a toupee, or cornrows, they should be removed before the appointment. No chemical treatments may be administered (coloring, perms, relaxers, etc.) within 48 hours before the QEEG. Hair must be free of beads, weaves, etc. The client should be instructed to bring a comb or brush.

4. Medications ~ The QEEG assessment is often cleaner and easier to read if there are no medications affecting the brain. If the client is taking stimulant medication (i.e., ADHD medication), it is preferable to do the QEEG recording after the client has stopped taking

the medication for up to 48 hours prior. The client MUST check with his or her prescribing physician or health care provider to determine if it is possible to stop taking the stimulants 48 hours prior to the QEEG. If 48 hours is not advisable, 12-24 hours is the next preferred length of time. Clients should not make changes in any other medication(s) unless authorized by their physician. If the client is taking medications for anxiety, depression, or sleep, they should NOT stop taking these medications without first consulting with his/her prescriber. If the prescriber approves, the client can bring these medications with them the day of the QEEG and take them after the QEEG assessment has been conducted.

5. Over-the-Counter Medications and Supplements ~ Unless prescribed by a physician or licensed health care provider, clients should avoid taking any over the counter medication or supplements for two or three days prior to the QEEG. This includes medications and supplements such as: acetaminophen (Tylenol), Advil (Motrin/ibuprofen), aspirin, analgesics, antihistamines/allergy medications (Benadryl, Claritin, Allegra, Zyrtec), cough and cold medicines, herbs, nasal sprays, nutraceuticals, sports drinks, (Gatorade, etc.), food supplements (including amino acids), vitamins, or other similar products.

6. Caffeinated Beverages ~ The client should NOT drink excessive amounts of coffee, tea, or caffeinated beverages on the morning of the testing (i.e., one cup is fine), and the client should NOT drink soft drinks with excessive amounts of caffeine in them (i.e., red bull, highly caffeinated soft drinks), for at least 15 hours prior to the QEEG. Some clinics ask that the client avoid all caffeinated drinks the morning of or 24 hours before the QEEG.

7. Alcohol and Drugs ~ Alcohol should be avoided 24 hours prior to the QEEG. Marijuana should be avoided 24-72 hours prior to the QEEG.

8. Contact Lenses ~ Portions of the QEEG require that the client's eyes be closed for up to 10 minutes. If the client wears contact lenses, he/she should be prepared to remove them if they create discomfort with closed eyes.

9. The client should bring a complete list of medications taken on a daily or regular basis the day of the QEEG.

10. Often, providers will acquire a mannequin head and put an EEG cap on it to show prospective clients what the capping procedure is like.

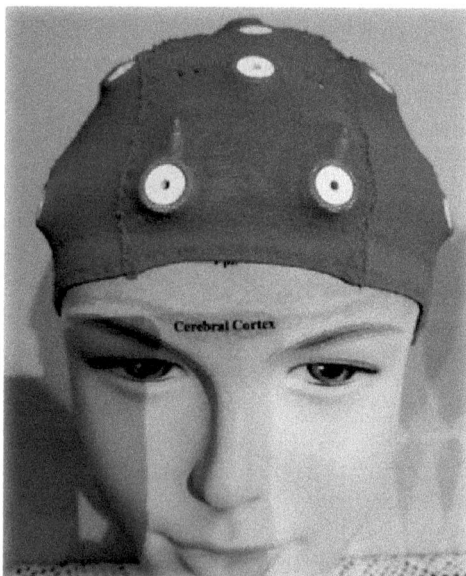

QEEG cap placed on mannequin head.

Day of QEEG Suggestions

The day of the QEEG, the client should:

1. Eat a high protein breakfast.

2. Women should not wear any makeup on the forehead or ear lobes.

3. Drink plenty of water the day before the QEEG recording.

4. Use the restroom prior to the start of the QEEG.

5. Do not wear jewelry on the neck or ears.

6. Nicotine should be avoided 3 hours prior to the QEEG.

7. Bring any medications or supplements you would like to take after your QEEG is complete.

On the day of the QEEG brain map appointment, the client should plan to spend a minimum of 60 minutes in the office and more time if the clinician plans to review the results of the QEEG and/or if the client needs time to change clothes, fix hair, etc.

Supplies Needed to Perform a QEEG

Not all seasoned clinicians prep and conduct QEEGs the same way. Most of us have developed individual styles, and you will too, over time. The recommendations below are standard for most clinicians conducting QEEGs.

One of the most popular QEEG caps for conducting QEEGs are the colored nylon caps sold by Electro-Cap International (see resource list at the back of this book). The following supplies are suggested:

1. Assorted QEEG Caps – Electro-Cap International makes caps of various sizes. Small

is yellow, medium is red, and blue is large. They are made out of nylon and stretch to fit over the head. They also make in-between sizes.

2. One Pair of Ear clip Electrodes

3. Head measuring Tape

4. NuPrep®

5. Electro-Gel®

6. Orange Sticks

7. 6" single end wooden Q-Tips (The wooden end slightly shaved to fit easily through the cap holes, but not sharp which would be uncomfortable for the client)

8. Two adhesive sponges for FP1, FP2 sites

9. China marker or pen

10. Single use syringes

11. Single use blunt needles

12. Small portable hair dryer

13. Two small hair clips to secure longer hair away from ears

14. Towel to prop behind client neck if needed

15. Towel for wet cap after it has been cleaned

Performing the QEEG

On the day of the QEEG, ask the client to come in a few minutes before the scheduled appointment time. This provides an opportunity for the client to fill out any paperwork (i.e., informed consent). Once the client is in the chair being prepped, is an opportune time to review some of the aspects about brain maps that were discussed during the consult or initial meeting as well as provide the opportunity for the client to ask questions. Prepping shouldn't take a lot of time (20-30 minutes). Measure the client's head for the right size cap, clean the forehead and earlobes with Nuprep, and place the cap on the head.

When choosing a cap size and placing on the clients' head, use the measuring tape from the nasion (notch at the top of the nose and slightly below the eyebrows) to the inion (bump on the back of the head that marks the bottom of the skull). Go up 10% and place a mark in the center of the forehead with a china marker or ballpoint pen. Do the same in the occipital area then measure around the head on these two marks with the colored head measuring tape, which indicates which size cap is appropriate. Using this circumference measure, calculate 5% and place a mark 5% distance on each side of the mark on the forehead, which is where you will place the FP1, FP2 sticky circles attached to the cap. For a rougher estimate, have the client look straight ahead and line up the two marks with their pupils.

It can be helpful to have two people place the cap on the client's head. One person can remove the paper covers and put the two sticky circles on the cap at FP1 and FP2 and place them on the dots on the client's forehead and then hold the front of the cap in place. The second person can gently pull the cap over the client's head while keeping the ear wires positioned correctly on either side and laying behind the client's ear. Slight cap adjustment can then be made to ensure the 10-20 sites are positioned correctly on the head, and no wires are pinched under the cap.

Once the cap is on the client's head, there are a few tricks that may be helpful in prepping the sites. Take the wooden end of a wooden Q-Tip and dip the tip into some NuPrep. Place the wooden end of the Q-Tip dipped in NuPrep into each of the electrode sites and gently use it to cleanse the scalp. Then add Electro-Gel to each site and once again use the wooden end of the Q-Tip to work the gel down to the scalp.

Another approach is that once the cap is on the client's head, we show the children the syringe filled with Nuprep and a second syringe of conductive gel.

We explain we are going to fill the ports, and it will be a slightly warm feeling (the Nuprep tube is left in the warmer overnight). We quietly show their parents the blunt needle and tap the top with our finger as we explain it is not sharp and simply used to fit through the ports. If the client is an adult, then we show the blunt needle to them in a similar fashion. We squirt in a small amount of Nuprep then use the back of a wooden Q-tip to work it down through the hair while filling the ports with conductive gel and checking impedance.

If possible, try to have the impedance down to 5 khm or less and within 1.5 khm across all sites. There are some clients where skin type or behavior issues will not allow that level, but do what you can and document the issue and level of impedance in your notes. Any readings under under 30 khm will give you good data. For younger clients it can be helpful to play a movie during the setup process to both help them sit still and to distract them from a potentially distressing process.

Once the client is hooked up and ready to go, have them sit where they can watch their brainwaves on the screen. Ask them to blink their eyes multiple times quickly and explain the pattern they see on the screen. Then have them roll their eyes and then have them bite down hard and grit their teeth. Clients are able to see how easily artifact is generated and how it negatively affects readings. Garbage in = garbage out!

When it comes time to record, we want the client's head in an upright, relaxed position with the eyes looking straight ahead. If their head is tipped down, you may see muscle artifact at O1 or O2, and if they lift their eyebrows, artifact can show up at FP1 or FP2. We generally mark a dime-sized dot on a sticky note and place it on the wall at eye level. We explain the positive and negative electrical impact if they move their eyes and encourage them to focus on the dot. We explain that it is normal for them to see color shifts and unusual spots as the brain is not used to the eyes staying still and plays tricks on them. We also encourage them to keep their tongue relaxed in their mouth and not touch their teeth together to avoid jaw tension.

Children, especially those who are hyperactive, can be very challenging. For younger children having them sit on a parent's lap can assist in keeping them still. For the eyes closed recording, the parent can place fingertips on the child's eyelids to help keep the eyes still. Using a sleeping mask can also be helpful. We have even had parents read or tell the child a story during the eyes closed measurements. For the eyes open recording, we have had the child watch a movie during the recording to keep them as still and focused as possible.

Cleaning up After the QEEG

After the QEEG is completed and you have verified and saved your data, the cap can be removed from the client's head. Once the cap is removed, alcohol prep pads can be used to remove any paste and or gel from the earlobes, and paper towels or make-up remover pads with rubbing alcohol can be used to wipe the excess Electro-Gel® off of the hair and scalp. Remember that cleaning up the client as much as possible gives them a more positive customer experience. They may not understand the complexity of the recording process, but they certainly understand if they leave your office with gel dripping off their hair versus leaving the office cleaned up and visibly put together.

If you are using an Electro-Cap®, it can be rinsed under warm or hot running water to remove the excess gel from the fabric part of the cap and wash away the gel inside the plastic electrode housing. While holding the ear lead connectors in one hand to prevent water from getting inside them, allow water to run through the cap electrode ports until it streams out the other end. Twirling the cotton end of a Q-tip in each port before a final rinse helps shift any large gel build up. This helps prevent buildup on the electrodes and electrode housing that can negatively affect future recordings.

Follow the manufacturer recommended cleaning instructions making sure it is clean and ready for use on the next client. One method to allow the cap to dry is to place it over an appropriately sized container such as a large candle jar with lid. Loop the ear lead connectors up through the bundle of wires running into the cap to make sure water does not drip into them and have a paper plate or towel under the jar. Once dry, the cap should be stored properly where it is ready for the next use.

Reviewing QEEG Results with Clients

There are multiple mapping software programs available to neurofeedback clinicians for running a QEEG, and among those reports, various types and amounts of information are provided to the clinician. As a clinician, it is important to understand the level of understanding of the client regarding brain map interpretation. Some mapping systems give reports that summarize findings, while others require the clinician to do all the interpretation. In either case, it is important to make sure that all information shared with the client is received at the client's level of understanding.

For some clients, just the process of having a QEEG conducted can raise anxiety, and for some hearing the results can lead to a range of emotions from anxiety to sadness to relief that there is proof of the issue and they weren't imagining symptoms. It is for these reasons we always encourage clinicians to have clients sign an informed consent before conducting a QEEG and/or starting neurofeedback.

Another important aspect of brain map interpretation is one's scope of practice. Mental health providers can make interpretations and/or recommendations regarding mental health disorders; however, they cannot make claims or treatment recommendations regarding a medical diagnosis. Therefore, to help the clinician provide the best information regarding QEEG brain map interpretation we recommend the following:

1. Always do a comprehensive intake and assessment of the client before conducting a brain map.

2. Do not use a QEEG brain map for making a diagnosis. A QEEG can be used ethically and professionally for purposes of assessment and protocol selection.

3. When reviewing the map and discussing findings such as sleep problems, anxiety, ADHD, etc., always check with the client to make sure the reported findings match symptoms the client has reported.

4. Be prepared to explain the differences that may occur between an eyes closed and eyes opened map.

5. Be prepared to explain differences when a client reports a particular problem or symptom and the brain map does not indicate abnormalities or deficits in the area of the brain where a particular symptom is likely to show up.

6. Be prepared to explain where neurofeedback training will occur and the reasons for training at specific 10-20 locations.

7. Have a specialized HIPAA release for clients to sign if you will be sharing results of a QEEG or requesting specific health information on the client from another practitioner.

We have a unique opportunity in our field to present detailed information to the client about their distressing symptoms and a tool that teaches their brain to change, thereby decreasing or eliminating their negative symptoms. Instead of sending the client home to grieve and make peace with their lot in life, we offer hope of change to their core brain state. Once training is complete, they have the capacity to move forward with far greater independence with their personal goals and desires.

Chapter Five

Neurofeedback Protocol Selection

Functions of the Brain

Before beginning neurofeedback, it is important to have a basic understanding of brain function. This includes not only basic anatomy but common pathways and connections. You should continue to build on this knowledge for as long as you are in the field.

Learning the International 10-20 system of electrode placement and Brodmann areas gives the clinician a good basic start to what areas of the brain perform what functions. Learning about the lobes of the brain and their respective functions will further that knowledge.

There are classes available at local colleges or online. BCIA coursework includes basic information as part of their certification process. Research is available from a variety of places such as libraries, ISNR and other membership organizations, and conferences.

Different Types of Neurofeedback Training

Traditional brain wave training/neurofeedback training started with single channel protocols (specific electrode placement training certain brain waves up or down), and slowly evolved to two channel training. That was all the equipment was able to do. By 2007 manufacturers were developing four channel devices, and within a few years amplifiers capable of doing 19 or more channels were being released.

Despite advancements in the field, much of the published research in neurofeedback was being conducted doing amplitude/magnitude training, and, as of the writing of this book, a major research project just released findings looking at single and two channel neurofeedback training.

Single channel and two channel protocols can be further broken down into monopolar and bipolar montages. A single channel monopolar protocol consists of one active electrode on the scalp with a reference electrode on an ear and a ground electrode on the other ear or on the scalp. A single channel bipolar protocol would have an active electrode on the scalp, a reference electrode on the scalp, and a ground electrode on the ear or scalp.

A two channel protocol would typically have two active electrodes on the scalp, reference electrodes on each ear, and a ground electrode; and a four channel protocol would typically have four electrodes on the scalp with a ground and reference electrode on each ear.

As the field has advanced, new systems have come out and some of those systems are capable of doing specialized neurofeedback training including coherence training, Z-score, and low-level currents that are delivered to the brain. Infraslow training and sLORETA training have become more popular over the past five years, and other advances continue to reach the market almost every year.

Neurofeedback Protocol Selection

Some QEEG brain map reports provide recommendations for particular protocols; however, the majority do not. In some cases, clinicians do not conduct brain maps and train clients based upon reported symptoms. Whether or not a brain map has been conducted, it is important to understand why the client is interested in neurofeedback and the chief problems/symptoms that need to be addressed.

Not all clients seeking neurofeedback training have problems or health-related symptoms for which they are requesting neurofeedback. Some persons seek neurofeedback for purposes of optimizing their personal performance at work or in a sport. Others are wanting to improve memory, and cognitive performance, and still others are seeking to improve upon a personal health task like meditation.

Protocol selection requires a basic working knowledge of the brain and brain function, what Brodmann areas—if any—should be targeted with neurofeedback, what sites should be trained, and what bandwidths are rewarded or enhanced vs. inhibited or stopped.

For some disorders, there are publications that review a particular symptom or disorder and specialized neurofeedback protocols that may be beneficial. Additionally, experienced neurofeedback practitioners often try, develop, and use protocols which may not be in the published literature. It is for these reasons we encourage the practicing clinician to become a member of ISNR in order to have access to membership through listservs where case questions, specialized protocols, and literature recommendations may be readily accessed. It is also common for a practitioner to have various mentors throughout their career as their interests and areas of specialty develop. These mentors can share their respective knowledge on a one to one basis and are often sources of both technical knowledge and emotional support.

Chapter Six

The Neurofeedback Session

The Room Set-up

Furniture

Typical room furniture and neurofeedback supplies vary. As a starting point, include a monitor, speakers, chair, and table with various supplies for the session. When purchasing a chair for an adult, keep in mind the client's comfort, whether you want armrests (we chose not to have them to avoid muscle artifact from various shoulder positions), and whether fabric or vinyl/leather works better with your clientele. Any cords or wires can be held together and hidden with Velcro straps made for this purpose. The picture below shows a child's chair, which is easily moved into and out of position depending on whether we are seeing an adult or child. We added carpet sliders to the chair legs to make movement easier.

Child's chair and moveable table for the amplifier

Miscellaneous Equipment

Consider white noise makers outside the training room to ensure privacy within and between offices. Electrical noise suppressors such as the Furman PST 2+6 Power Station, which musicians use to eliminate noise interference from power lines or other appliances plugged into the same circuit, can be extremely useful in eliminating electrical interference from your session.

Amplifiers are placed in a variety of locations, depending on the type of amplifier and arrangement of the room. Some clinicians have a cloth pocket on the back of the client chair for smaller amplifiers. Other clinicians place the amplifier on a small end table, which is easily moved near the client when running sessions and set out of the way when not in use (See above picture with child's chair and end table). Velcro pads on the table and amplifier decrease the risk of the amplifier getting bumped or moved while sitting on the table. Other clinicians have their amplifier in one location and have longer connecting wires or leads running to the client.

Supplies

Neurofeedback supplies

Typical supplies are shown in the picture above. From left to right is a jar of 10-20 conductive gel with a clean wooden stick used to place gel on a one-use client plate. Avoid any contamination issues by putting 10-20 gel and Nuprep on the one-use plate rather than scooping 10-20 gel from the jar with the leads. Next is the square container with precut cotton squares (Pro-Shot cleaning patches), which come in various sizes and shapes and go over the top of the leads, helping them adhere to the scalp. Torn pieces of Kleenex or cotton balls are also used for this purpose. Next is a container with makeup remover pads. We tried a number of brands to find one that was thick enough to hold the rubbing alcohol but did not disintegrate while scrubbing the scalp. This makes the clean-up process faster and more thorough.

Next is the bottle of rubbing alcohol with a pop-up lid and pump-action distribution. The next container is a business card holder with PDI scrubbing pads and following that the 6-inch wooden Q-tips that allow you to part the hair with the wooden end and rub the Nuprep on the scalp site with the covered end. This should be thrown out after it has been used on one client.

The stack of plates in the photo is set-up for the next client and will be thrown out after each session as each plate is single-use. Behind the plates is a warmer (similar to those used for ultrasound gel), which warms the Nuprep. We place it on the plate right before working with the client so that when we scrub the ears, the Nuprep is still warm. Finally, a Kleenex box is next to the warmer for wiping off the leads after the session before a more thorough cleaning with rubbing alcohol or boiling water. As with the make-up remover pads, try different brands as some flake and make a mess on your table where others hold together far better.

Remember that the client may not understand the complexity of what you do as a clinician, but the elements they personally experience as part of the process are important. For instance, leaving the office with a clean scalp rather than one with excess conductive gel left in the hair allows them to go about their day as normal after they leave. Making it obvious that you are keeping common supplies (e.g. conductive gel) separate can set their mind at ease that you are aware of contamination issues. Give your clients good customer service in ways they personally experience.

Client Orientation to Neurofeedback[1]

One of the most important tasks in starting up a new client in neurofeedback is orienting that person to the neurofeedback process. This often decreases their stress level and offers them an opportunity to ask clarifying questions. Your client is a new customer who may not have any background in your field. The more you communicate what you are doing and why you are doing it before you do it, the less likely they will be concerned or have a negative experience.

In the ISNR Learning Series published by The Foundation for Neurofeedback and Neuromodulation Research (FNNR), the book Becoming Certified in Neurofeedback: A Guide to the Neurofeedback Mentoring Process for Mentors and Mentees (2020) by R. Longo & R. Soutar, provides an overview of the requirements necessary to become Board Certified. One of those requirements is a review of the Essential Skills List, established by BCIA, to assure that practitioners properly orient new clients to the neurofeedback process.

Client orientation should include the following:

1. In layman's language, explain to a new client EEG biofeedback, self-regulation concepts, and operant conditioning of brainwave activity.

---

1   For more details on this topic see A Consumer's Guide to QEEG Brain Mapping and Neurofeedback Training by Robert Longo (2018).

2. Explain the major stages in the neurofeedback training process, from initial intake and assessment to progress monitoring and reporting.

3. Explain client's role and responsibilities in the neurofeedback process.

4. At initial session, explain how the neurofeedback session process and equipment works, including:

   a. purpose and steps involved in skin preparation

   b. steps in electrode attachment and selection of site placements; assure client

   c. about safety of "sensors"/electrodes

   d. meaning of primary features of the feedback screens and concepts of amplitude and

   e. frequency and/or z-scores

   f. relationship between client activity and on-screen feedback changes

   g session recording and progress monitoring screens.

5. Obtain written client permission for treatment/ training using a thorough Informed Consent form.

New clients should be properly introduced to neurofeedback and related areas. As noted above, all clients should be required to sign an informed consent before beginning neurofeedback training. In addition, clients should be aware of the costs, time requirements, and related health information prior to beginning neurofeedback services (see sample informed consent document in the Forms section in the back of this book).

Many neurofeedback practitioners, both seasoned and novices, maintain a list of seasoned professionals whom they may contact for mentoring and/or consultation when faced with a tough case. Many practitioners are specialized in treating common yet specific disorders, i.e., anxiety, depression, or cognitive functioning deficits. Others have experience in treating more complex disorders, i.e., Parkinson's, seizures, stroke, tic disorders, pain, addictions, etc.

If you have not been mentored or are looking for a mentor, *see Becoming Certified in Neurofeedback: A Guide to the Neurofeedback Mentoring Process for Mentors and Mentees* (2020) by R. Longo & R. Soutar.

Risks, Benefits, Costs, Time and Logistics of Sessions

Core and basic areas that should be addressed before beginning neurofeedback services should include but are not limited to the following:

1. Risks and benefits of neurofeedback (addressed in the informed consent)
2. Number of sessions and cost per session
3. Frequency of sessions
4. Side effects of neurofeedback
5. Health-related factors that improve the quality of neurofeedback training (sleep hygiene, diet, exercise)
6. Factors that may negatively influence neurofeedback benefits
7. Use of medications and transitioning off medications
8. Progress tracking
9. Eyes closed vs. eyes opened training
10. Do's and don'ts before and after

a neurofeedback session.

11. Use of alcohol and recreational drugs

Preparing clients for neurofeedback sessions can be unnerving at first but with time and practice, it will become as routine as brushing your teeth. Some clients have few questions, and others have done reading about neurofeedback and will have more challenging questions.

We believe it is important for the clinician to provide as much information up front as possible to best prepare the client for neurofeedback sessions.

Common Client Questions

1. How long is each session?

Sessions will vary in length based upon age, protocol, and clinician preference. On average, most sessions are 20-30 minutes.

2. Will I feel anything during or after the session.

Most clients do not feel anything during the sessions with traditional neurofeedback, though a few will mention they feel tired by the end of the session. We let clients know that their brain is working very hard, and they may feel extra tired for the first few sessions until their brain becomes accustomed to the training. This process is similar to going to the gym, where you build up endurance and muscle.

3. How many sessions do I need?

We generally explain that we offer information to their brain but cannot predict how quickly it will respond and shift. We do tell them that on average our clients do two to three sets of fifteen sessions. This number will vary depending on the type of neurofeedback used, the client's issues, and whether they are trying to wean off medications. Some clients have responded well after 20 sessions, while others with severe disorders may take 50 or more. As with other information in this book, adjust your answer to the specific particulars of your business.

4. What changes should I be looking for?

There are changes that will be specific to each client's situation and this is a good time to give additional information around how neurofeedback is a gradual learning process (operant conditioning), and so they can expect to see an improvement that builds over the course of training. This is why we periodically check in with symptom sheets to track what they observe. Our symptom sheet covers a range of general categories, and then we often ask client specific follow-up questions based on the issues they initially cited when they came in for training.

5. What does eyes closed training entail?

Eyes closed training typically involves training with eyes closed and listening to some form of auditory reinforcement, i.e. a sound, tone, or music with volume fluctuation based upon training parameters.

6. What does eyes opened training entail?

Eyes open training typically involves a visual stimulus that varies the brightness of an educational DVD or streaming of an educational show. Some practitioners use video games while others have the client watch a video while coaching them.

It is also advisable to prepare clients for sessions by discussing time of day, what to eat or drink before a session, what to expect after a session, and any homework or health care advice you wish the client to follow.

Running the Neurofeedback Session

Common Clinician Questions.

For the clinician, there are several items you should understand and be aware of in conducting neurofeedback training sessions.

1. Is it okay to leave the client alone during sessions?

This depends on your software and clinic policy. We encourage a philosophy that once a client enters the training room, we do not leave it. While running the session, the neurofeedback provider watches the raw brain wave activity and manually adjusts the settings to ensure that optimal training occurs and no equipment issues have cropped up. The clinician also provides verbal feedback if a client needs to relax facial muscles, shoulder muscles, or calm physical movement.

We prefer that the clinician or technician stay in the room with the client as one needs to be present in case there is a software problem or technical issue as shown below.

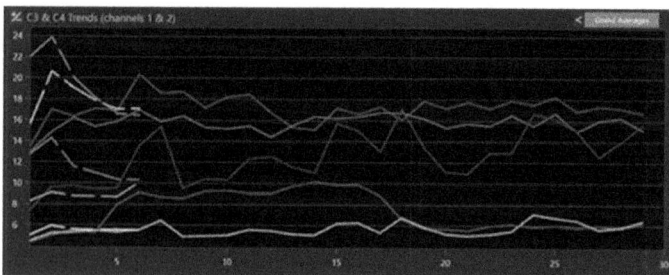

Notice how the signal on one channel stopped. This happened because an electrode came loose.

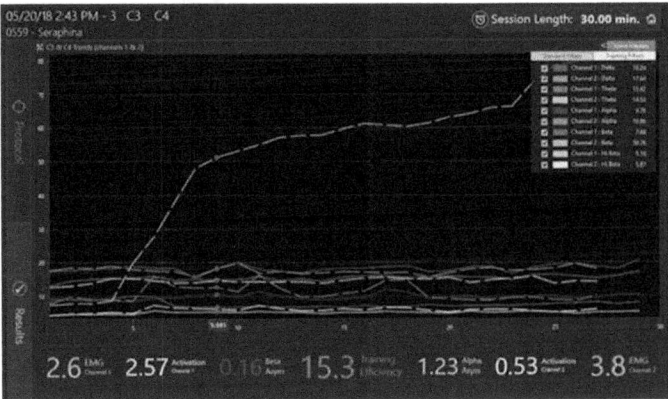

The pink dotted line (Beta) went up when the client's cell phone went off.

> 2. What should I do if the client falls asleep during a session?

If the client is training with eyes closed and falls asleep, they will train equally as well if they fall asleep.

If the client is training with eyes open and falls asleep, pause the session, let the person take a short 2-3 minute cat-nap, then wake them up. Once they are awake and alert, resume the session.

> 3. How do I know if the session is running well?

As you learn more about neurofeedback and the equipment you are using, you will see sessions during which the client does not have a good training session and others where the training session went well. In general, the power should be near or within normal ranges, and the individual brain waves (Delta, Theta, Alpha, and Beta) are condensed (close together). If conducting a two channel training, the EEG wavelengths from each hemisphere are close and integrating.

Different equipment may use different software, and trend screens will vary in how they look as in the diagrams below. With some software, you can look at an after-session graph and note whether the amplitude you are inhibiting decreases and the coherence you are rewarding improves. We periodically show this graph to clients while reminding them that the QEEG will give us more precise information, but the graph shows us trends across time.

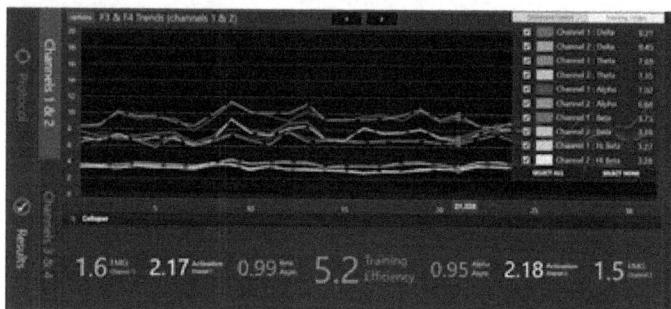

In this example, the trend screen shows EEG that is condensed and integrating.

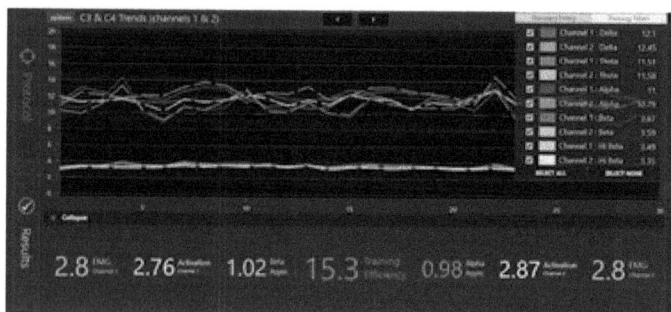

In this example, all of the slow waves (Delta, Theta, and Alpha) are excessively high.

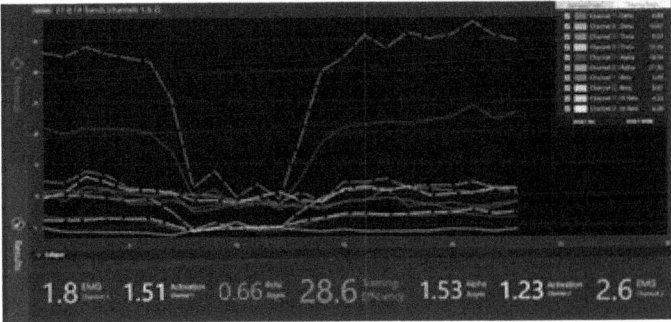

In this example, the Alpha waves are excessively high. This is the sign of a busy mind where the client is thinking excessively versus trying to relax and remain focused. The person succeeds briefly in reducing the Alpha, but soon returns to a busy mind.

4. How do I determine if the client is making progress?

Improvement in training sessions is one indicator that the client is doing well. The example above of a good training session is one indicator when the sessions are consistent. Asking the client, their parent or significant other, and a teacher or tutor to fill out symptom sheets on a consistent basis (every 3-4 sessions) is another method of determining overall progress. Additionally, repeat QEEGs and cognitive testing can help a clinician gather data showing progress.

5. What should I do if the client begins to have a negative side effect?

Some neurofeedback sessions may result in negative side effects. Common side effects may include mild irritability, mild headache, or mild anxiety. If the client reports a negative side effect, but it goes away within an hour or two of the session, then the clinician may want to continue with the same protocol. If, however, the client has a severe reaction or side effect, changing protocols may be in order.

Protocols based on QEEG results are less likely to result in negative symptoms because they are targeting specific areas of the brain that are out of line. Protocols based solely on client-related symptoms are more prone to guesswork and can result in more negative symptoms.

6. Explain to the client what they should be doing during the session (i.e., what rewards they should expect, the frequency, etc.)

During any neurofeedback session, the client's focus should be on paying attention to rewards and how the person is able to achieve the most rewards. In an eyes open session, if watching a DVD for example, the goal would be to stay focused and keep the viewing screen bright. During an eyes closed session, the focus might be on listening to the tones and keeping them coming once every 1-2 seconds. In the example above where the client had a busy mind, the rewards would be less frequent. Thus, the overall goal is to stay relaxed and focused on the session and to avoid thinking about "things" that can cause distractions and even a negative emotional state.

Introducing the Client to the Equipment

A brief introduction to the equipment should have occurred during the introductory meeting or intake assessment. When the client arrives for their first session, show them the leads or cap that is used, the Nuprep and conductive gel, and how the leads connect to the amplifier and computer. Some clients will want to see their raw brain wave activity on the clinician's screen, and others simply want to do the session without any extra information. In each case, adjust the information you offer to your client according to their age and interest level. Remember that as the client moves through training, they will often continue to ask questions. We build in a few minutes in each session for this specific purpose and tell the client or their parents so they are aware that we are available and interested in answering their questions and concerns. This availability often allows us to gather information about client progress and offer additional resources depending on their needs.

Explaining the Set-up Process

Similar to the QEEG set-up, we show the client what we are doing before we do it to avoid any concerns or anxiety. Children sometimes want to touch the Nuprep or gel and become interested in the process rather than frightened. We give parents the option of staying in the training room during the entire training process or waiting in the waiting room. They often join their child in the training room during the first few sessions and then stay in the waiting room after that.

Applying Leads and Running a Session

We clean the client's ears with a PDI pad then scrub with a wooden Q-tip with Nuprep on it. Then the ear lead with 10-20 conductive gel is applied. Specific 10-20 sites are found, and those sites scrubbed with a Q-tip dipped in Nuprep before each of four leads laden with 10-20 conductive gel is applied, and a precut cotton fabric square (pro-shot cleaning patches) applied to help hold it in place. We check impedance, and if any sites are above the standard, they are scrubbed and rechecked. Occasionally there are times when a child has sensitivities or behaviors, and it is better to have a slightly higher impedance and ensure the client can receive the training. For the most part, an attentive job with set-up ensures an appropriate impedance level.

This set-up process will be slightly different depending on the type of training used. However, the core principles of preparing the ears and head, correct lead placement, and impedance levels remain the same.

Running the neurofeedback session is also specific to the type of equipment and software you have purchased. Some software programs have very specific behind the scenes settings. Be sure to read the manual, access tutorials, work with a mentor, and reach out for customer support if you have any questions. The equipment is only as good as the correct set-up and run of the hardware and software.

Tips and Tricks

Many neurofeedback practitioners conduct neurofeedback in addition to therapy sessions, while others provide strictly neurofeedback services and make referrals for therapy should a particular client need additional interventions. In either case, you should establish a routine that accomplishes everything you need to do for/with the client in order to maximize your time. Some clinicians are busy enough to hire and train a neurofeedback technician to work within the practice while others do prep, sessions, and clean up themselves.

Some tips for maximizing your time include:

1. Have spare sets of electrodes available to you in case an electrode goes bad. The same goes for other electronic cords (e.g., HDMI or power cords). Keep spare sets of batteries handy in the session room if your amplifier/impedance meters run on battery.

2. Clean the electrodes between each client. One of the quickest methods it to heat up water in a cup to boiling or near boiling (microwaves are good in this regard as are individual serving hot water heaters). You can dip the electrodes into the water, and swish them around, cleaning the leads of paste and sterilizing the electrodes. If using cupped ear-clips, open the ear clip spring

to separate the ear-clips and swish them in the hot water. Another method is to soak a makeup remover pad in rubbing alcohol and wipe down the electrodes and wires.

3. Set up your supplies (electrodes, cotton gauze, 10-20 Paste, NuPrep, Alcohol Prep pads, etc.) on a table or tray, so they are ready when the client comes in and sits in the chair.

4. If you are doing eyes closed training and incorporate music into sessions, have the music set up and ready to play. If you are using DVDs for eyes opened training, have the DVD ready to insert into the DVD player.

5. Have disposable wipes available to clean surfaces between clients. If you see children, a small vinyl or leather chair is easily wiped down with disposal cleaning wipes and sprays are available for cloth covered chairs.

6. Have charts or notes pulled up and create templates for forms you typically use.

Chapter Seven

Reviewing Neurofeedback Sessions with Clients

Reviewing Session Progress

There are different types of neurofeedback (traditional 1, 2, and 4 Channel training, sLORETA, Z-Score, Infra Slow, LENS, etc.), and different brands of equipment, some of which are limited in the types of neurofeedback they can run; and the various trend/session screens produced by the software that run the sessions.

You should make sure you have been adequately trained in the equipment and software you use so that you can comfortably explain what you show the client about their session if they ask to see the training screen.

You should be able to explain what bandwidths you are training up or down, why you are doing so, and the results you are hoping to achieve within a certain number of sessions.

Reviewing Symptom Changes

We always ask the client how they felt during and after the session, and what symptom improvement they may have noticed, i.e., feeling calmer, more relaxed, more alert, better focus, etc. You can also have the client fill out a one-page symptom sheet every 3 to 4 sessions (examples are included in the next chapter and the appendices) rating their progress on a scale from 1-10 in various areas. An area for them to write additional information is included.

Clients should also be asked to report any negative side effects during or after a training session. It is possible for a client to experience negative side effects. The use of a QEEG in protocol selection greatly decreases the likelihood of negative side effects.

In general, neurofeedback will not interfere with most other treatments. Neurofeedback has few side effects when administered properly. The most common side effects of neurofeedback include improved sleep, more awareness of dreams, feeling calmer, feeling more energy, and feeling more focused. Temporary side effects such as headaches, insomnia, anxiety, feeling giddy, agitated, or irritated may occur during or right after a neurofeedback session; however, these side effects can be adjusted and eliminated immediately in most cases. It is also possible that clients might fall asleep during or after neurofeedback sessions.

When do you Discuss Continuation or Completion?

Each practice is different, and the specific number of sessions between repeat assessments/QEEGs depends on a clinician's experience, the type of clients they see, and their method of measuring progress. Questions you should ask yourself as you make these decisions are:

1. How do you demonstrate and show progress?

2. How many sessions will you do between completion check-in or repeat testing?

3. How do you balance the need to prove results and collect data for clinical direction and research versus client costs?

4. How often do you do repeat QEEGs?

In our clinic, we typically run fifteen sessions of neurofeedback training. Between session twelve and thirteen we have a more in-depth conversation with the client or their parents as to whether they are pleased with the level of negative symptom reduction and want to finish training or they want to continue.

If they are ready to graduate, we do not do a repeat QEEG as we do not feel it is necessary for the client to incur the additional cost when they are happy with their current level of function. If they want to continue with another set of training we proceed forward with repeat testing. This repeat testing shows progress the client has made—which should be shared with the client—and gives us information we use to determine the next training protocol. If we were doing a research project, then the concluding testing would be essential, and the question becomes whether the clinician covers this cost or asks the client to cover this expense.

Chapter Eight

Session Documentation and Record Keeping

Client Documentation

There are several items that clinicians often use to document contact/sessions with clients. Intake forms, assessments, and session notes are very common. Some neurofeedback/QEEG mapping systems provide assessments, neurofeedback session data, and built-in forms that can be filled out online and saved and/or downloaded and printed out.

As noted earlier, taking a comprehensive history during the intake process is important. In the field of neurofeedback there are many key areas that are important to assess such as levels of anxiety, levels of depression, quality of sleep, overall health, personality factors, cognitive processing, and emotional processing. There are a number of different ways to assess these issues initially and throughout the training process. CNS Vital Signs has a light cognitive assessment that is commonly used by neurofeedback providers, as does Cambridge Brain Sciences.

Session Documentation and Record Keeping

Depending on the practice you work in, progress notes may be handwritten, electronic, or a combination of both. In the forms section of this booklet are examples of forms one might use to document neurofeedback sessions. In many instances, practices and clinics use the SOAP or DAP format for documentation.

Session notes should include the following: pre-post session measure of EEG and specifically the EEG of areas being trained, type of neurofeedback and protocol(s) being used, sites where training occurred, length of session, and type of feedback (audio/ visual, etc.). The use of additional modalities such as (Biofeedback, Photic) should be noted, as well as client response to session, any negative side effects that may have occurred, and observed changes in client physical posture, movement, facial expression or verbal interaction.

Client Progress or Symptom Tracking

Tracking client progress is important. In general, one should see some improvements in areas like sleep within the first few sessions. Some QEEG mapping systems include built-in client progress tracking options. Whether you use such a system or not, creating a progress tracking method for each client is important. We recommend you limit symptoms that clients will track to 10 or 12 items. A written checklist with a rating system the client can use to fill out each week will be useful for tracking progress.

The example below is a simple method to track progress:

Name _____ Date _____

Think of each item as a PROBLEM and rate each using the following: 1: No, minimal, or never a problem. 2-3: Occasional or mild problem. 4-5-6: Sometimes, moderate problem. 7-8-9: More often, more bothersome, significant problem. 10: Always or very severe problem.

Concentration   _____
Short Term Memory   _____
Quality of Sleep   _____
Appetite   _____
Motivation/Energy   _____
Positive Moods   _____
Patience   _____
Assertiveness   _____
Awareness of Dreams   _____
Relaxed/Calm   _____

Restlessness   _____
Worry/Negative Thinking   _____
Negative Mood*   _____
Negative Emotions   _____
Depression   _____
Fatigue   _____
Irritability   _____
Anxiety/Stress   _____
Anger   _____

The following is another example of a symptom tracking sheet where the client circles a number on a scale from 1-10 and has the option of writing in additional information.

Neurofeedback Training Report

Name_____Date_____

You may notice some of the following changes as a result of the training. Please rate any noticeable changes **since beginning neurofeedback**. Five is neutral, and ten shows the greatest improvement.

	←worse better→
1. Trouble sleeping/sleeping better	1 2 3 4 5 6 7 8 9 10
2. Less energy/more energy	1 2 3 4 5 6 7 8 9 10
3. Anxious, nervous/Calm, relaxed	1 2 3 4 5 6 7 8 9 10
4. Poor concentration/better concentration	1 2 3 4 5 6 7 8 9 10
5. Sad, down/ happier, feeling up	1 2 3 4 5 6 7 8 9 10
6. Spacey, foggy/ more awake, alert	1 2 3 4 5 6 7 8 9 10
7. Felt physically worse/ felt physically better	1 2 3 4 5 6 7 8 9 10
8. Change in social interaction worse/better	1 2 3 4 5 6 7 8 9 10
9. Change in social conversation skills worse/better	1 2 3 4 5 6 7 8 9 10
10. Any changes in medication since your last visit?	No Yes
11. Any major changes in your environment since your last visit? (Includes moving, remodeling, diet, relationships, jobs, schools etc.)	No Yes

12. Describe in the space below any examples of changes you've observed since the last session:

Chapter Nine

Case Summary and Closure

Case Summary and Closure

There is no general rule for determining when a client has successfully completed neurofeedback. For most mental health disorders such as anxiety, depression, etc., the conventional wisdom has dictated 30-40 sessions. With advances in equipment and software and increased knowledge in running sessions, ISNR now indicates a range of 20 to 40 sessions. For persons with severe disorders such as a traumatic brain injury, the number of sessions can reach into the 50s or 60s— or more.

There are several philosophies around ending neurofeedback training. One method we recommend is to run about 15 sessions and see how the client is doing. At that point, many clients are beginning to notice significant improvement, and session benefits are lasting week to week. If the client is coming in for two or more sessions per week, then reduce the number of sessions by one session per week every two weeks until the client is down to one session per week.

If the client reports no slippage or reduction in symptom improvement and the benefits and changes are holding, then the clinician can slowly begin to reduce frequency of sessions. Once down to one session per week and after 20 sessions, try skipping a week. If the benefits hold without negative change or resumption of symptoms, schedule the next session for two weeks. After the second two-week period, if the client reports doing well and no resumption of symptoms, schedule the client for the next session three weeks out. If after three weeks the client reports doing well, no slippage, no resumption of symptoms, then consider termination of neurofeedback sessions at that time.

Other clinics report using an approach where they start with the intake assessment, run a set of 15 sessions, evaluate progress using a repeat QEEG and cognitive test, and run another set of 15 sessions. At the end of each set of 15 sessions, the clinician and client discuss symptom improvement and whether they want to continue or graduate. If they continue, the QEEG and cognitive evaluations are done, and training continues. If they are satisfied with their symptom improvement, they stop training at the end of their set of 15 sessions.

A written summary or documentation of the session reduction process, client self-report, and progress tracker information can then be written and placed in the client file and neurofeedback sessions terminated. If the client requests a more formal written summary, the information will be readily available in the documentation. Summary reports to other providers can also be generated from this information.

Graduation Gifts

We give a graduation gift that contains a number of the client's favorite items in a large basket wrapped in cellophane. We allocate ~2 - 4% of the total amount spent during training. During sessions we subtly ask for favorite foods, drinks, toys, and other interests and try to include them in the gift basket. The gift baskets are unexpected and memorable, and increase the feelings of goodwill toward us. Clients and parents are more inclined to return for additional training later, and especially refer their friends and acquaintances when they leave us on such a high note.

If you include marketing materials in the graduation gift, you may be able to deduct the cost of the gift as a marketing expense on your taxes.

Chapter Ten

Equipment Maintenance

Hardware and Software Maintenance

In general, neurofeedback equipment is well made and long-lasting. Keeping it in a safe location where it can't be dropped or easily damaged is all one needs to do. If and when you have an equipment problem or failure, contact the manufacturer for specifics as to how to best address the problem.

Software, on the other hand, can be more problematic. Software can be corrupted and sometimes software updates can be problematic in making all of the different programs talk to one another as they should. If you are experiencing a software issue, often the problem can be resolved by simply closing the software and rebooting the computer. If that doesn't fix the problem, contact the software manufacturer for proper guidance and maintenance. Many clinicians have chosen to have a designated neurofeedback computer which is not connected to the internet to avoid automatic updates, which may interfere with neurofeedback software function.

Cleaning

If you are working in a medical facility you may have to follow the guidelines established by infection control. We would recommend the following. Client chairs should be leather, vinyl, or a smooth cleanable surface so that they can be easily wiped down on a regular basis—especially between clients. Cloth chairs are more difficult, time-consuming, and expensive to clean and much more likely to carry germs.

Computer keyboards, mouse, etc., can be wiped down

with disinfectant wipes. QEEG caps can be cleaned with soap and water or germicides. It is important to follow the manufacturers' suggestions for cleaning. Electrodes can be cleaned with an alcohol wipe or simply dipped into boiling water for both cleaning and sterilization.

Troubleshooting

Several problems can arise before or during a neurofeedback session, and it is important to become familiar with troubleshooting and resolving these problems. Issues might arise with software, discussed briefly above. In some cases, the issues may be equipment-related, arising from a bad electrode or a bad wire somewhere in the session leads or in the QEEG cap. Excessive and constant artifact and/or poor impedance readings are common occurrences of electrode problems.

Another common source of problems is electrical interference. Electrical interference with neurofeedback equipment can be caused by a variety of factors: equipment plugged into an electrical outlet not properly grounded, a client sitting close to an electrical outlet, a cell phone on their person, a microwave being turned on nearby, multiple phone lines running through the office, Wi-Fi devices, and even a nearby vending machine dispensing a soft drink.

The photos below are examples of electrical interference problems.

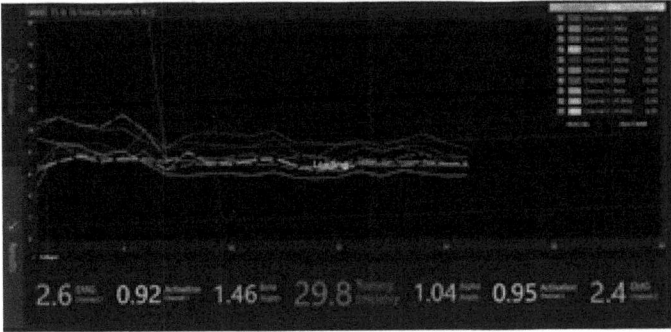

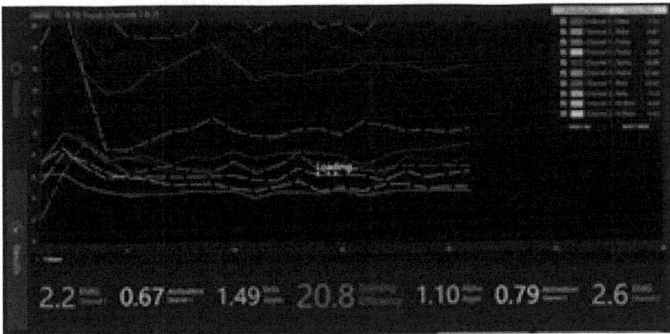

The sudden drop in Beta (pink line) in the above photos occurred after the clinician unplugged the power pack to the computer, which was causing electrical interference.

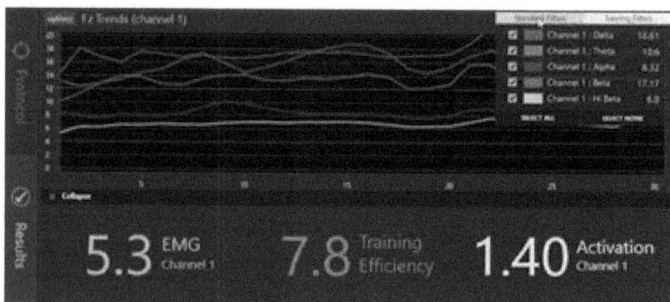

The sudden elevation in Beta (pink line) above occurred as the result of electrical interference during the neurofeedback session.

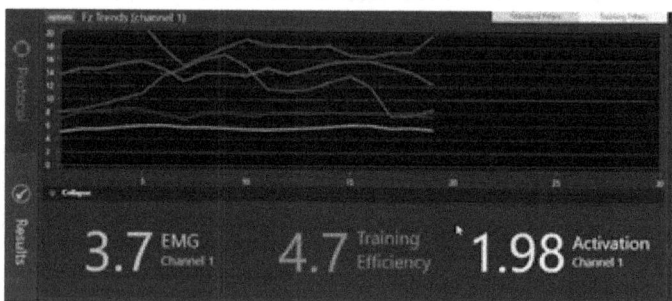

The sudden rise and drop in Beta (pink line) above occurred as the result of electrical interference during the neurofeedback session. Excessive (electro-magnetic fields (EMF) and dirty electricity, (i.e., ungrounded electrical outlets, ungrounded computer plugs, cell phone, nearness to Wi-Fi routers, can create this type of electrical interference.

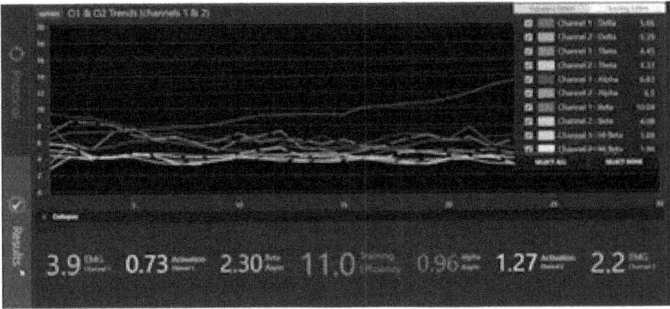

The steady rise in elevation in Beta (pink line) above occurred as the result of electrical interference during the neurofeedback session.

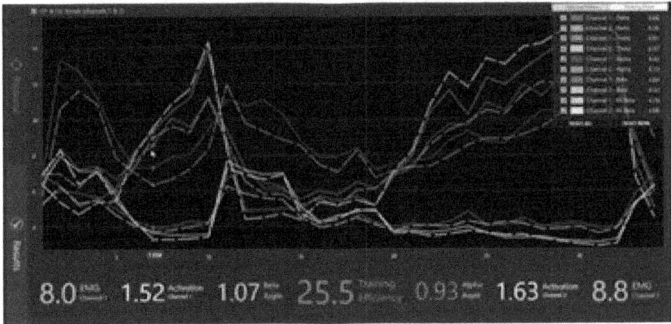

Excessive movement by the client while sitting in a chair can create trend screens that look like the one above.

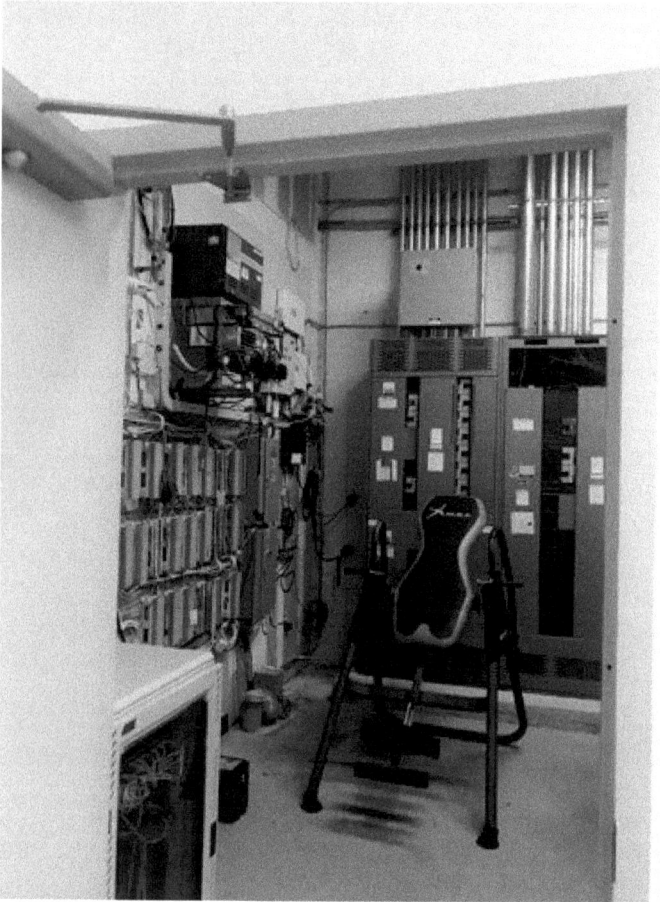

The equipment closet in this office building housing high power electrical equipment such as X-Ray machines, CAT Scans, MRI, fMRI, etc. can create electrical interference if it is close to your office and neurofeedback equipment.

High power electric lines can affect neurofeedback sessions due to electrical interference.

Replacing Leads

Most electrodes will last quite a while. They are typically made of various metals, including tin, silver, or a base metal with gold plating. We find that the gold-plated electrodes are not as long-lasting as silver and tin as the gold plating wears off, exposing the underlying metal. A list of companies that sell neurofeedback related equipment and supplies are listed in the appendices.

A simple way to test electrodes to see if they are working is to plug them into the neurofeedback device and stack them together (see photo below). Pull up a session and run it with the electrodes stacked. This shorts the electrical system out, and one will get smooth lines with very low magnitude readings. If you disconnect one, you will get a bad signal (poor impedance and poor EEG readings. So, as long as the EEG signals are clean and smooth (see photo below), the electrode is good. If you have a bad electrode, you will have a poor EEG signal (see photo below).

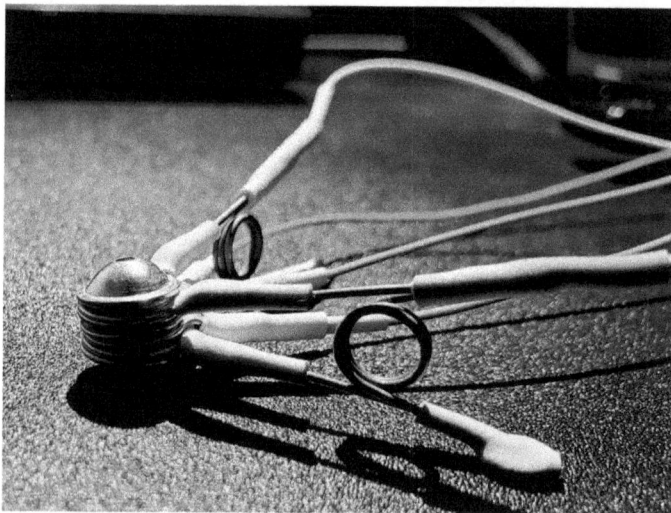

Stacked electrodes. Place one electrode on top of the other and hold together with ear clips.

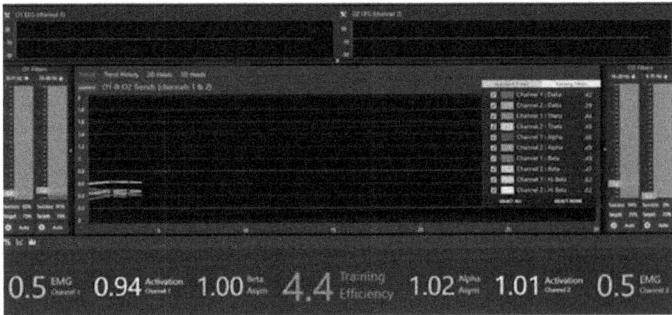

At the top of the trend screen are two clean EEG signals, one for channel 1 and the other for channel 2.

In this example, one electrode (Channel 2) has been removed from the stack. The EEG signal is poor. If you have a bad electrode when all the electrodes are stacked, you'll get a bad signal. If it is the ground, both EEG signals will be bad.

Conclusion

Neurofeedback is a highly specialized intervention that requires targeted training, mentoring, equipment, software, and commitment to continuous learning. It is an amazing tool that can have a profoundly positive impact on the brain. The improvement clients can experience as their negative symptoms decrease or are eliminated has tremendous positive implications for their lives, the lives of their family, and society as a whole. In our opinion, it is well worth the effort. We are excited to have you enter the field, both for your own growth in an area that will constantly challenge you and for the benefit of the clients whose lives you will change for the better.

Appendices

Partial List of Manufacturers of Neurofeedback Equipment

We do not want to show any bias for or against equipment manufacturers and suppliers. This list is not all inclusive but was created from a list of venders who have been sponsors/venders at ISNR conferences over the past few years or who offer neurofeedback/QEEG-related supplies. They are listed in alphabetical order. Each has a website where their products can be reviewed.

Amazon - amazon.com

Applied Neurosciences – appliedneuroscience.com

BrainMaster – brainmaaster.com

BrainPaint – brainpaint.com

Bio-Medical—bio-medical.com

Deymed – deymed.com

EEG Education and Research –eeger.com

EEG Sales – eegsales.com

Electro-Cap International – electro-cap.com

LENS (Ochslabs) – ochslabs.com

NewMind Technologies – newmindtraining.com/

Neurofield – neurofield.com

Pro-Shot (cloth squares for lead application) proshotproducts.com

Thought Technology – thoughttechnology.com

Weaver and Company – weaverandcompany.com

Forms

PROGRESS NOTE (Sample Form) #1

Business Name

Client's Name: ID#:   Date of service:

Subjective (client/family report and/ or clinical interview data):

Objective: Psychophysiological monitoring using EEG.

Observations:

Impressions/assessment/comments:

Plan: Continue to use one, two, or four channel neurofeedback magnitude training and monitor for progress in self-regulation skills.

Multimedia: Run Length: xx Minutes

Notes/Comments:

Session # 1   Total Sessions: 1

_____

Therapist's Signature

PROGRESS NOTE (Sample Form) #2

Health and Behavior Assessment /Intervention

Client's Name: Date of service:

Subjective (client/family report and/or clinical interview data):
Objective:

Psychophysiological monitoring or procedures:

Heart rate variability sensor location:

Peripheral temperature sensor location(s):

Thermofeedback:

EEG sensor location (s):

Respiration:

Other psychophysiological monitoring:

Health oriented questionnaires:
Cognitive/Behavioral or other mind/body interventions:
Observations:
Impressions/assessment/comments:
Plan:

_____

Therapist's Signature

PROGRESS NOTE (Sample Form) #3

EEG Biofeedback Session Data Sheet

Name: _____   ID#___LI0054___   Protocol name/ID:
Sites: Threshold:
Run Length:
Training Protocol:

Date _____   Session #_____/_____
Clinician_____

	Delta	Theta	Alpha	LoBeta	Beta	HiBeta	9-11Hz
Channel 1 Pre							
Channel 1 Post							

Comments

	Delta	Theta	Alpha	LoBeta	Beta	HiBeta	9-11Hz
Channel 1 Pre							
Channel 1 Post							

Comments:

Comments:

At the beginning of each session, please fill out this checklist to help us track and evaluate your progress. On a scale of 1 – 10 regarding each of the items below.

		Worse				Average	

Better
Item
1(Lo)   2   3   4   5   6   7   8   9   10
(Hi)

Concentration   _____
Short Term Memory   _____
Quality of Sleep   _____
Appetite   _____
Motivation/Energy   _____
Positive Moods   _____

Patience	_____
Assertiveness	_____
Awareness of Dreams	_____
Relaxed/Calm	_____
Restlessness	_____
Worry/Negative Thinking	_____
Negative Mood*	_____
Negative Emotions	_____
Depression	_____
Fatigue	_____
Irritability	_____
Anxiety/Stress	_____
Anger	_____

* An emotion lasts for 20 minutes to an hour, a mood lasts for several hours, days, or weeks

COMMENTS:

PROGRESS NOTE (Sample Form) #4

Neurofeedback Training Report

Name_____ Date_____

You may notice some of the following changes as a result of the training. Please rate any noticeable changes since beginning neurofeedback. Five is neutral, and ten shows the greatest improvement.

	←worse better→
1. Trouble sleeping/sleeping better	1 2 3 4 5 6 7 8 9 10
2. Less energy/more energy	1 2 3 4 5 6 7 8 9 10
3. Anxious, nervous/Calm, relaxed	1 2 3 4 5 6 7 8 9 10
4. Poor concentration/better concentration	1 2 3 4 5 6 7 8 9 10
5. Sad, down/ happier, feeling up	1 2 3 4 5 6 7 8 9 10
6. Spacey, foggy/ more awake, alert	1 2 3 4 5 6 7 8 9 10
7. Felt physically worse/ felt physically better	1 2 3 4 5 6 7 8 9 10
8. Change in social interaction worse/better	1 2 3 4 5 6 7 8 9 10
9. Change in social conversation skills worse/better	1 2 3 4 5 6 7 8 9 10
10. Any changes in medication since your last visit?	No Yes
11. Any major changes in your environment since your last visit? (Includes moving, remodeling, diet, relationships, jobs, schools etc)	No Yes

12. Describe in the space below any examples of changes you've observed since the last session:

Sample Informed Consent

PLEASE READ AND SIGN BELOW

The purpose of this form is to obtain your voluntary consent to participate in one or more methods of Quantitative Electroencephalography (QEEG) Brain Mapping, Peripheral Biofeedback, Neurofeedback, other forms of relaxation and stress reduction interventions, and to disclose potential benefits and risks associated with these interventions. (Business name) provides various educational interventions, assessment protocols, and health care services, a few of which are still considered, by some, to be experimental.

QEEG Brain Mapping

In order to determine an appropriate Neurofeedback training plan, a QEEG performed by _____, using the -_____ expert referential database system will need to be conducted.

(Business name) will assess your need for having a QEEG. In order to engage in neurofeedback, you will be required to have a QEEG assessment. In other instances, to help verify a disorder, your doctor, or another health care professional, may recommend you have a QEEG. A QEEG consists of placing a cap on your head with 19 electrodes/sensors. Each site will be cleansed and a special gel will be placed under each sensor to insure proper conductivity to read your brainwaves. Preparation and the assessment procedure take approximately one hour.

Benefits: QEEG may help me further understand and/or confirm the problems/symptoms, disorders, and/or diagnosis for which I am seeking assessment and health care services.

Side Effects/Risks: QEEG may result in my feeling anxious/apprehensive, and/or uncomfortable during the procedure, and sad/disappointed regarding findings from the procedure. The cap may cause you to have a mild headache.

Forensic Services: QEEG Brain Mapping for purposes of neurofeedback is not a medical procedure and is not done at (business name) for purposes of medical diagnosis. Data collected is not done in a manner that meets the Daubert criteria for admissibility of evidence in court. (Business name) does not provide forensic services or diagnosis for TBI. We do not accept invitations for depositions. Those seeking a diagnosis for TBI or any other medical or mental disorder should seek services of a medical physician or a forensic neuropsychologist. Your signature below indicates you agree not to request or seek such services from us presently or in the future, or through third parties such as legal counsel or insurance companies.

Client Rights. You have the right to:

1. Decide not to receive QEEG Brain Mapping services from us. If you wish, we can provide you with the names of other qualified QEEG providers.

2. End the QEEG at any time.

3. Ask questions about protocol and procedures used during the QEEG procedure, and to ask questions about QEEG techniques if you feel unsure of them.

4. Have all that you say treated confidentially and be informed of state law placing limitations on confidentiality in the QEEG relationship. Under certain circumstances, we are required by law to reveal information obtained during a QEEG assessment to other persons or agencies without your permission. Also, we are not required to inform you of our actions in this regard. These situations are as follows: (a) If you threaten bodily harm or death to yourself or another person, we are required by law to notify the victim and appropriate law enforcement agencies; (b) If a court of law issues a subpoena; (c) If you are having a

QEEG or being tested by a court of law, the results of the QEEG assessment must be revealed to the court; (d) If you have given us information concerning non-accidental injury and neglect to minors or incompetent adults. (e) If you are in the process of filing a workman's compensation claim or file such in the future.

Equipment/Software: QEEG measures will involve the use of the (type of equipment) software and hardware (type of equipment). (Type of equipment) products are FDA registered. QEEG maps are produced using (type of software/databases).

Neurofeedback Training

Neurofeedback involves several electrodes/sensors being placed on the scalp and earlobes. The sensors detect brain wave activity including Alpha, Beta, Delta, and Theta brainwaves. Individual brainwaves are measured and revealed on a computer screen revealing your brainwave activity. Through instruction you can learn to train down or train up certain brainwaves associated with stress management, attentional, cognitive and/or emotional deficits and related disorders. In some cases, neurofeedback must be considered as experimental. Trainings last from 10-30 minutes and may occur two or more times per week for an average of 25-30, and in some cases 30 or more sessions.

Benefits: Neurofeedback is known to assist individuals by decreasing symptoms associated with brain and central nervous system dysfunction. Other benefits include the possibility of reducing problem behaviors and increasing peak performance. In many cases, neurofeedback is considered to be experimental when used to a train certain disorders. Please feel free to ask for a more detailed explanation regarding your particular problem area or training interest.

Side Effects / Risks: Neurofeedback will not interfere with most other treatments. Neurofeedback has few side effects when administered properly. The most common side effects of neurofeedback include improved sleep, more awareness of dreams, feeling calmer, feeling more energy, and feeling more focused. Temporary side effects such as headaches, insomnia, anxiety, feeling giddy, agitated, or irritated may occur during or right after a neurofeedback session; however, these side effects can be adjusted and eliminated immediately in most cases. It is also possible that you might fall asleep during or after neurofeedback sessions.

Client Rights. You have the right to:

1. Decide not to receive Neurofeedback services from us. If you wish, we can provide you with the names of other qualified Neurofeedback providers.

2. End Neurofeedback sessions at any time.

3. Ask questions about protocol and procedures used during Neurofeedback training, and to ask questions about techniques if you feel unsure of them.

4. Have all that you say treated confidentially and be informed of state law placing limitations on confidentiality in the Neurofeedback relationship. Under certain circumstances, we are required by law to reveal information obtained during training to other persons or agencies without your permission. Also, we are not required to inform you of our actions in this regard. These situations are as follows: (a) If you threaten bodily harm or death to yourself or another person, we are required by law to notify the victim and appropriate law enforcement agencies; (b) If a court of law issues a subpoena; (c) If you are being trained with Neurofeedback, at the direction of an attorney or Medical doctor for legal purposes, the results of the training or tests must be revealed to the court; (d) If you

have given us information concerning non-accidental injury and neglect to minors or incompetent adults. (e) If you are in the process of filing a workman's compensation claim or file such in the future.

Equipment/Software: Neurofeedback training will involve the use of (type of equipment) software and hardware. (type of equipment) products are FDA registered.(repeat of previous info)

Other Methods: Other treatment methods may not work as rapidly as the methods and modalities described above. Alternative methods of treatment and/or therapy include traditional medical treatments, medications, the use of supplements, the use of relaxation techniques, group and individual therapy.

Choosing the Right Intervention: The interventions described above are voluntary, not mandatory. You will not be pressured for not participating. You may withdraw from/stop receiving Neurofeedback training sessions at any time without consequence.

Consent

I voluntarily consent to participate in and undergo the assessment and/or intervention methods and modalities described above. I understand that I am free to withdraw my consent and to discontinue participation in the interventions/modalities/methods described above at any time. The natural consequences and potential risks and benefits have been fully explained to me by (business name).

Permission

My signature below indicates that I have read, reviewed and understand this informed consent (and/or I have had the form and its contents read to me and explained to me), and I consent to participate in the procedures described above. I understand I may ask questions at any time and may request to stop interventions at any time. I have read and understand my rights.

Signed:_____ **Date:**_____

Printed name:_____

              (FIRST)   **(MIDDLE)**   **(LAST)**

If Minor: Parent or Guardian_____

Date_____

If you have an emergency after regular business hours call 911, or contact your personal physician

NEUROFEEDBACK ESSENTIAL SKILLS LIST

A beginning neurofeedback practitioner should be able to demonstrate mastery of the following basic skills, as attested by their BCIA-approved Mentor who will initial each item as completed.

Client Orientation

1. In layman's language, explain to a new client EEG biofeedback, self-regulation concepts, and operant conditioning of brainwave activity.

2. Explain the major stages in the neurofeedback treatment/training process, from initial intake and assessment to progress monitoring and reporting.

3. Explain client's role and responsibilities in the neurofeedback process.

4. At initial session, explain how the neurofeedback session process and equipment works, including:

a. purpose and steps involved in skin preparation

b. steps in electrode attachment and selection of site placements; assure client

c. about safety of "sensors"/electrodes

d. meaning of primary features of the feedback screens and concepts of amplitude and

e. frequency and/or z-scores

f. relationship between client activity and on-screen feedback changes

g. session recording and progress monitoring screens.

5. Obtain written client permission for treatment/training using a thorough Informed Consent form.

Intake, Assessment and Protocol Selection

1. Document a thorough client symptom and medication history and gather background information relevant to treatment/training goals

2. Provide a thorough EEG baseline assessment, using the following skills:

a. perform correct measurements to name and locate on the scalp each of the International 10-20 System electrode placement sites

b. properly prepare scalp and ears and attach electrodes to selected assessment sites or attach an electrode cap if doing a full-cap quantitative EEG

c. correctly perform all steps to collect a QEEG recording or multi-channel EEG

d. assessment: checking impedances, removing artifact, and collecting eyes-open and eyes-closed data

e. demonstrate basic understanding of a QEEG assessment report, as well as the most commonly reported components of QEEG databases (absolute power, relative power, phase, coherence, z-score comparisons, etc.)

f. identify recordings indicating spike and wave activity requiring consultation with a neurologist or QEEG expert

g. use all intake, psychometric, and baseline EEG assessment data to select target electrode placement sites and montages for neurofeedback treatment/training

h. select an initial neurofeedback protocol and explain rationale to client.

Use and Maintenance of Neurofeedback Equipment

1. Demonstrate thorough knowledge of operation

of neurofeedback equipment of choice:

a. Make correct hardware connections
   and start hardware.

b. Make correct electrode connections
   to the hardware.

c. Identify and remove (or control for) sources
   of common artifacts in the raw EEGsignal.

d. Troubleshoot common equipment failures
   according to manufacturer's recommendations.

2. Demonstrate   thorough   knowledge   of
   appropriate software for selected equipment:

a. Accurately select, install, and run
   neurofeedback treatment/training software.

b. Identify components, applications, and
   limitations of selected software package.

Neurofeedback Session Management and Reporting

1. Conduct   neurofeedback   treatment/training
   sessions involving the following procedures:

a. Provide initial orientation and instructions to
   client at first treatment/training session.

b. Prior to subsequent sessions, query client
   (and/or parent) verbally and/or via pre-

c. Session questionnaire on client's positive and
   negative reactions to previous session.

d. Maintain basic hygiene procedures in
   attaching (and cleaning) electrodes.

e. Remind client of the training objectives
   for session and their role in attending
   to and responding to feedback.

f. Start treatment/training software program,
   set up selected protocol parameters,
   and run basic feedback functions.

g. As appropriate, set initial training thresholds and adjust as needed.

h. Identify and remove sources of artifact appearing in session recordings.

i. Monitor session recordings and provide coaching and supplemental verbal feedback to client during sessions, as appropriate.

j. Save session data per equipment guidelines and review session results with client.

k. Assign homework to client that supports and supplements session training goals.

j. Consult with client's prescribing physician and/or providers of other concurrent treatments as necessary to avoid treatment complications and maximize treatment outcomes.

l. Identify as soon as possible in the treatment/training process when neurofeedback is not working for a client; identify cause(s) for lack of progress; make necessary protocol or other training program adjustments; or, when necessary, recommend termination of neurofeedback.

m. In collaboration with client, determine when neurofeedback treatment/training goals have been met and mutually plan for treatment termination and follow-up.

n. Conduct all aspects of neurofeedback treatment and training in accordance with BCIA, AAPB and ISNR codes of ethical practice.

2. Maintain orderly and up-to-date client files, including

a. session-by-session training records, significant session events and client comments

b. changes in client medication, significant life changes, allergies, etc. that may impact treatment/training results

c. reports of consultations with other treatment providers, family members, teachers, etc.

Use of Supplemental Therapeutic and Training Modalities

1. Demonstrate ability to establish positive, constructive relationships with clients and their family members, using basic counseling and communication skills

2. Document adequate training and demonstrate skills required to use appropriate counseling/ therapy methods to supplement neurofeedback with clients having mental health diagnoses

3. Document adequate training in use of alpha-theta neurofeedback protocols. Demonstrate ability to select appropriate clients for alpha-theta training as well as apply appropriate therapy methods when using these protocols

4. Document adequate training in other neuromodulation modalities (such as HRV biofeedback, AVS, CES, etc.) for use in conjunction with neurofeedback, and demonstrate ability to select and use appropriate adjunctive modalities with individual clients.

References

AddictionBlog.org. (2011). How long does marijuana, weed, pot (THC) stay in your system? Retrieved June 22, 2018, from http://drug.addictionblog. org/how-long-does-marijuana-weed-pot-thc-stay-in-your-system

Biofeedback Certification International Alliance. (2020). *Neurofeedback Certification.* Biofeedback Certification International Alliance.https://www. bcia.org/i4a/pages/index.cfm?pageid=3431

Chen, J. (2020, May 5). Social media demographics to inform your brand's strategy in 2020. *Sprout Social.* https://sproutsocial.com/insights/new-social-media-demographics/

Coben, R., Arns, M., & Kouijzer, M. E. J. (2011). Enduring effects of neurofeedback in children. In *Neurofeedback and neuromodulation techniques and applications* (pp. 403–422). Elsevier Inc.

Coben, R., Wright, E. K., Decker, S., & Morgan, T. (2015). Impact of Coherence Neurofeedback on Reading Delays in Learning Disabled Children: A Randomized Controlled Study. *NeuroRegulation*, *2*(4), 168–178. https://doi.org/10.15540/ nr.2.4.168

Hirshberg, L., Chu, S., & Frazier, J. A. (2005). Emerging interventions. *Child and Adolescent Psychiatric Clinics*, *14*(1), 1–19. https://doi.org/10.1016/j. chc.2004.07.011

International Society for Neuroregulation & Research. (2020). *Recommended Reading.* ISNR; globalneurofeedback. https://isnr.org/ recommended-reading

Levinthal, C. F. (2016). Drugs, behavior, and modern society(Eighth). Boston, MA: Pearson.

Longo, R.E. & Soutar, R. (2020). *A guide to the neurofeedback mentoring process for mentors and mentees.* Greenville, SC. FNNR.

Longo, R.E. & Soutar, R. (2019). *Becoming certified in neurofeedback.* Greenville, SC. FNNR.

Longo, R.E. (2018). A consumer's guide to understanding QEEG brain mapping and neurofeedback training. iUniverse, Bloomington, IN.

Misner, I. (2018). The Five Types of Business Networking Organizations [Web log post]. Retrieved June 23, 2019, from https://ivanmisner.com/five-types-business-networking-organizations/

Morrison, K. (2014, November 28). *81% of shoppers conduct online research before buying [infographic].* https://www.adweek.com/digital/81-shoppers-conduct-online-research-making-purchase-infographic/

Sherlin, L. (2016). *Introduction to QEEG.* https://vimeo.com/ondemand/novatecheeg

Steiner, N. J., Frenette, E. C., Rene, K. M., Brennan, R. T., & Perrin, E. C. (2014). In-school neurofeedback training for ADHD: Sustained improvements from a randomized control trial. *Pediatrics, 133*(3), 483–492. https://doi.org/10.1542/peds.2013-2059

Volkow, N. D., Swanson, J. M., Evins, A. E., DeLisi, L. E., Meier, M. H., Gonzalez, R., ... Baler, R. (2016). Effects of cannabis use on human behavior, including cognition, motivation, and psychosis: A review. JAMA Psychiatry, 73(3), 292. https://doi.org/10.1001/jamapsychiatry.2015.3278

Lightning Source UK Ltd.
Milton Keynes UK
UKHW020754031022
409835UK00011B/1413